In Context

History and the History of Technology

Research in Technology Studies

In Context

History and the History of Technology

Essays in Honor of Melvin Kranzberg

Research in Technology Studies,
Volume 1

EDITED BY
Stephen H. Cutcliffe
AND
Robert C. Post

Bethlehem: Lehigh University Press
London and Toronto: Associated University Presses

Associated University Presses
440 Forsgate Drive
Cranbury, NJ 08512

Associated University Presses
25 Sicilian Avenue
London WC1A 2QH, England

Associated University Presses
P. O. Box 488, Port Credit
Mississauga, Ontario
Canada L5G 4M2

The paper used in this publication meets the requirements of the American National Standard for Permanence of Paper for Printed Library Materials Z39.48-1984.

Library of Congress Cataloging-in-Publication Data

In context.

(Research in technology studies ; v. 1)
"A select bibliography of the publications of Melvin Kranzberg": p.
Bibliography: p.
1. Technology—History.   2. Kranzberg, Melvin.
I. Kranzberg, Melvin.   II. Cutcliffe, Stephen H.
III. Post, Robert C.   IV. Series.
T15.15   1989   609   87-45969
ISBN 0-934223-03-3 (alk. paper)

PRINTED IN THE UNITED STATES OF AMERICA

Contents

6 CONTENTS

Illustrations

Melvin Kranzberg

Foreword

STEVEN L. GOLDMAN
STEPHEN H. CUTCLIFFE

This volume inaugurates the series, *Research in Technology Studies,* to be published by Lehigh University Press under our general editorship. The objective of the series is to present a representative range of technology studies scholarship bearing on issues engaging the community of scholars studying science, technology, and society relationships. Each volume in the series will have a topical theme and will consist of essays invited by the guest editor(s) of that volume.

It is singularly appropriate that the inaugural volume of *Research in Technology Studies* should be dedicated to Mel Kranzberg. The existence of an academic technology studies community in the United States is in large measure a result of his life's dedication to that end. The essays in this volume reflect the continuing strength of the history of technology as a pillar of technology studies scholarship even as the field expands to embrace established disciplines from philosophy and policy analysis to engineering studies and management science. The essays reflect as well the standards of scholarship that Kranzberg established for the young discipline almost thirty years ago with the founding of what is still its premier journal, *Technology and Culture.*

The aim of *Research in Technology Studies* is to carry forward in yet another, explicitly interdisciplinary, way the enterprise launched so conspicuously by Mel Kranzberg, an enterprise that has become increasingly momentous as the significance for society of science and technology, and the complexity of the mutual relationships among scientific research, technological innovation, and social values, have begun to be appreciated.

Introduction

STEPHEN H. CUTCLIFFE
ROBERT C. POST

Technology is today recognized as an integral component of social change. Furthermore, it is increasingly acknowledged that technology cannot be understood outside its social context. Historians, and especially historians of technology, have come to recognize the role that cultural, political, and economic values have played in shaping technological innovation, as well as the role that technological innovation has played in shaping values. Such a "contextual" appreciation of technology is part of a larger effort to understand and control the interactions of technology and culture. Hence, a better understanding of our technological past can contribute to the practical end of demystifying technology and helping us plan for the sort of futures we want to realize.

In the scholarly study of the history of technology, as the field has evolved over the past three decades, one individual stands in a unique position, Melvin Kranzberg. Mel Kranzberg's enthusiasm, his organizational skills, his editorial sensiblities, and his intellectual leadership have propelled the Society for the History of Technology (SHOT) and its quarterly journal, *Technology and Culture (T&C)*, to the forefront of what has become a major area of scholarly inquiry and discourse, both in the United States and internationally. For these reasons, and because of the central role that the history of technology has come to play within the wider field of technology studies, this volume honors Mel Kranzberg by offering a collection of essays that reflects on the history of technology as a discipline—on its past, its current state, and its potential. If indeed the history of technology has become a major scholarly specialty, three decades ago no such future could have been predicted. Long perceived as poor relatives among historians of science, historians of technology labored alone, or nearly so, in ascetic cloisters of academic obscurity (if, indeed, they had any academic standing at all).[1] Following a failed attempt to obtain some greater standing within the History of Science Society, Mel Kranzberg and a small, well-chosen cohort set forth to establish a new society and a new journal.[2]

Despite some initial difficulties, both enterprises succeeded. They succeeded, perhaps, even beyond Mel Kranzberg's dreams, but in any case the fundamental determinant of that success was surely Mel's enthusiasm and dedication. In his dual role as secretary to the society and editor-in-chief of its journal, he pressed established scholars to contribute to the nascent field and at the same time encouraged younger historians to direct their energies to the history of technology as an emerging specialty whose direction they could help shape while making their own professional mark.

Through voluminous correspondence, and always with the superlative supportive touch that would become a Mel Kranzberg hallmark, he spread the word and enhanced the scholarly reputation both of SHOT and of its journal, eventually building the one into the leading society of its kind in the world and the other into the leading journal. In the process, Mel himself clearly emerged as the primary force behind a significant new discipline. Because of his unique role in shaping that discipline, all historians of technology can think of themselves as students of Melvin Kranzberg.

Other organizations and journals, to be sure, were concerned about technological devices and processes—the Newcomen Society, for example. But Mel Kranzberg's SHOT was avowedly concerned "not only with the history of technological devices and processes but also with the relations of technology to science, politics, social change, the arts and humanities, and economics." Clearly, this breadth of scope and purpose has helped the society remain open to new perspectives. At the same time, because SHOT's founders were very conscious about being innovators, and their successors have kept this same consciousness, the field has always had a propensity for self-examination.

Over the years various scholars have inquired into the essence of the field and analyzed the history of technology's position vis à vis other ancillary specialties as well as history more generally. Conferences such as the one on "Critical Issues in the History of Technology" held in 1978 in Roanoke, Virginia, have sought to define questions that ought to be central to the discipline. Most recently, John Staudenmaier's thought-provoking analysis of T&C's first twenty years, Technology's Storytellers, has raised our collective consciousness of the field's attainments and its shortcomings, and caused us to consider its future direction. So it has been within an established self-reflective tradition[3] that we have conceived and drawn together this collection of essays.

As the history of technology has emerged as a major academic discipline, it has been shaped by a host of productive scholars who have analyzed particular technologies, usually within a set of broader cultural concerns. But there is a subset of that scholarly community made up of people who have reflected on the history of technology as a field of intellectual inquiry, a field unique in some regards but at the same time sharing much with other fields. Because this book is essentially about the field of the history of technology rather than about the full range of its specific concerns, it was among these latter scholars that we sought contributors.

While anxious to avoid too narrow a set of constraints, we did ask each contributor to focus on some theme that was reflective of the field's historical development and whose thrust was fundamentally historiographical. As with any book of this sort, there is considerable diversity. Some essays address subjects of longstanding concern, such as the roots of the Industrial Revolution or the relationship between science and technology. Some analyze the relationship of the history of technology to other historical fields or to nonhistorical disciplines, such as the philosophy of technology. Some consider the utility and value of particular bodies of source material. Some address the philosophical precepts and commitments of individuals who have had a major impact on the field. Common to all the essays, however, is a sense of the importance of a contextual approach, both historical and historiographical—a shared perception that, we believe, endows the volume with a substantial degree of thematic coherence.

* * *

The nature of technological knowledge and its relationship to scientific knowledge is a recurrent concern among historians of technology. Ever since the publication of his 1971 *T&C* article, "Mirror Image Twins: The Communities of Science and Technology," Edwin T. Layton, Jr., has been at the forefront of this discussion.[4] It seems entirely fitting, then, that his 1986 SHOT presidential address lead off the essays in this book. Layton commences with a delightful parody of National Public Radio's "Prairie Home Companion" and Lewis Carroll's *Through the Looking Glass,* in which the rooms on either side of the glass are labeled "science" and "technology," then goes on to suggest that science and technology can best be understood through use of an "interactive model." Neither one can be adequately addressed without reference to the other. The relationship, Layton concludes, is "symbiotic, egalitarian, and interactive," and both science and technology need to be evaluated within the appropriate social context and in consideration of the goals and values of each community.

The Industrial Revolution is another perennial interest among historians of technology, and in the second essay Robert P. Multhauf analyzes that watershed event in a new light. He finds British industrialization to have been especially strongly influenced by Dutch "engineers"—that is, "civil" engineers as distinct from military engineers. The authors of Renaissance "machine books" had generally been called "engineers," apparently after Vitruvius's master craftsmen who were possessed of "ingenium." The transfer to Britain of the technical knowledge contained in these Continental machine books was in large part occasioned by the immigration of Dutch engineers such as Humphrey Bradley, who became involved in such projects as harbor construction at Dover and the draining of the Fens. Multhauf concludes by asking, "Is it not likely that Dutch custom made the British mechanic an engineer—and the steam engine an engine instead of a machine?"

If the history of industrialization is explicated in part by an analysis of its terminology, it is also the case that much is revealed about the history of technology by the editorial directions taken in its technical journals. Eugene Ferguson presents several examples of the influence editors have had on the course of technology. *Scientific American* promoted those devices whose inventors were clients of its associated patent agency. In contrast, Zerah Colburn of *Engineering,* who had no such personal axes to grind, publicized a wide range of topics, both English and American, while *Engineering News* under the editorship of George Frost emphasized the educational value of analyzing technological failures. Finally, Ferguson shows how Matthias Forney, editor of *Railroad Gazette,* used his journal to counter the extravagent claims of partisans of the "narrow gauge." Ferguson concludes that, although technical journals may be valuable source materials, historians need to be aware of their editorial agendas—"what hobby or obsession or loyalty may stand behind the campaigns and crusades. . . ."

Although a far different genre from technical journals, various late nineteenth–century entertainments provide an equally revealing body of sources. Bruce Sinclair describes, for example, the celebratory theme of technological progress presented in a popular ballet, *Excelsior.* Throughout a series of historical vignettes depicting Papin's conception of the steamboat, Volta's invention of the battery and subsequent developments in telegraphy, the building of the Suez Canal, and the tunneling of the Alps, enlightened technology emerges triumphant. International expositions also documented technological progress. Particularly apposite examples were the series of expositions organized in Paris by the French engineer Frederic Le Play, in which even the buildings were designed to convey a sense of the relationship between technology and social advance. Sinclair shows that a study of cultural artifacts can provide a significant measure of a society's attitudes towards technology.

It has taken a long time for historians to rid themselves of the misconception that technology somehow evolves under its own imperatives, separate from human direction and cultural context. Yet, as Arthur P. Molella shows, three pioneers were fostering a sophisticated approach to the history of technology two generations ago. Abbott Payson Usher saw history as composed of "systems of events," separable lines of causally linked developments among which technology was particularly dynamic and thus especially worthy of analysis. Lewis Mumford evidenced a profound ambivalence about the cultural ramifications of "technics," at once fearful of the deleterious effects of mechanization and hopeful about how it might be directed toward humane ends. Sigfried Giedion, by means of an ingenious examination of the societal essence of mundane technical artifacts and processes, sought to alert humankind to its capacity to control its technological destiny. All three members of this "first generation," as Molella terms it, viewed technology not mechanistically but in holistic terms that recognized the interdependent relationship of technology and culture.

Like Molella, Thomas Parke Hughes returns to the works of Giedion and Mumford and, with them, that of Jacques Ellul. Hughes shares with all three of these men the view that the United States is best characterized as a nation of "machines, megamachines, and systems." In Giedion's view, mechanization applied to inappropriate domains posed a threat to the emotional side of human personality. Mumford addressed the large-scale rationalization of various resources into nonorganic technological systems. That theme would subsequently be echoed by Ellul, for whom humans had become a mere technical component in a vast technological system. If the nation has become fundamentally technological as Ellul proposes—and Hughes suggests that this is in fact the case—then it follows that those forces we need to understand in order to control our destiny "are not now primarily natural or even political, but technological." Implicit in Hughes's analysis is an expression of hope that historians of technology will be able to contribute important insights as we seek to understand ourselves and the technological world we have created.

The fact that scholars such as Giedion, Mumford, and Ellul were philosophers as well as historians (or even primarily philosophers) provides a natural transition to Paul Durbin's essay on the emergence of the philosophy of technology as a field in its own right, and on that field's relationship to the history of technology. Durbin begins by noting that the built-in tensions between abstraction-oriented philosophers and detail-oriented historians are often balanced by significant complementarities. The works of historian John Staudenmaier and philosopher Carl Mitcham are expressive of important methodological similarities. Durbin proceeds to analyze Thomas P. Hughes's *Networks of Power,* concluding that Hughes fails to refute completely the philosophical notion of technology as an "autonomous" force, but at the same time noting that philosophers could come to a better understanding of the complex realities of technological development by studying Hughes's book. Durbin asserts that excellent history, as typified by David Noble's *America by Design,* can be written in support of a strong philosophical stance, and that subjects such as artificial intelligence in turn offer philosophers an excellent opportunity to address technological specifics as well as abstractions. Finally, he considers his own work on the social disruption engendered by high-technology societies, concluding that it would be impossible for him to make a credible argument for liberal reform if he could not use evidence carefully gathered by historians.

Durbin's work on "the relationship between Delaware's R&D community . . . and the state's traditional cultural institutions" of course falls within an approach increasingly prevalent among historians of technology, the approach dubbed contextual by John Staudenmaier. As Merritt Roe Smith and Steven C. Reber point out, however, there are different varieties of contextualism. Many historians still focus primarily on the technical, bringing in cultural factors "chiefly as background against which the main theme of technical development unfolds." Others locate the dynamism on the social side, casting the technical factors as backdrop. To suggest the range of this spectrum, Smith and Reber analyze four recent books by Thomas P. Hughes, David Hounshell, David F. Noble, and Ruth Schwartz Cowan.[5] Hughes, they find, regards technology primarily as "expanding knowledge," and Hounshell likewise emphasizes technical development, albeit with some greater attention to its social implications. With Noble the emphasis clearly shifts to social factors, the technical being deemphasized, and Cowan's analysis is framed in terms of "social, rather than technical dynamics," with technology "largely an independent variable." Work such as Noble's and Cowan's, because it deals with issues that are "complex and sensitive," poses a threat to older historiographical traditions and, yet, is ultimately beneficial insofar as it works to break down barriers tending to separate history of technology from general history. This point is later addressed by Brooke Hindle, but with a quite different assessment of its relative costs and benefits.

Expanding his own conceptual framework of contextualism, John Staudenmaier uses automotive technology to explore the three stages of design, momentum, and senility which he previously identified as inherent in a technological life-cycle, and the three constituencies of design, maintenance, and impact, which he associated with any given technology. What Staudenmaier finds is that the development of a specific technology, the automobile, was part of "a much larger social process by which the nation adopted a distinctive technological style." The concomitants of this larger process included mass production, modern advertising techniques, and a system of "standardization" at the expense of "negotiation." A diminished capacity for negotiation, Staudenmaier fears, is too high a price to pay for the benefits of standardization—of which there are admittedly many. He concludes that to understand our technological society, and to enable us to take charge of its direction, we must ascertain the roles played by all the constituencies who have a stake in successful technologies.

One of the most important facets of the contextual approach to emerge in recent years is the analysis of gender and the role of women in technological change. Judith McGaw and Joan Rothschild have been at the forefront of this analysis. McGaw admonishes historians to be more sensitive to the ways in which assumptions about gender are "intricately interwoven into the fabric of our culture." She insists that gender is simply an ideology, not a biological or behavioral reality. Ruth Cowan has clearly demonstrated the degree to which women are active shapers of their own destiny, yet some scholars persist in treating women as passive victims. Because men and women are constrained by similar social and economic circumstances, we ought to study masculinity as well as femininity and attempt, as McGaw concludes, to understand both halves of the ideology. As a key to broad areas of historical understanding, we must move beyond the notion of "separate spheres"—for example, the household and mechanized mill—as "logical" units of analysis. Questioning instead the functions of "this socially constrained separation" is certain to further our understanding of American industrialization. That sort of analysis, McGaw argues, will lead to a better integration of the history of technology into general history.

In her essay, Joan Rothschild elaborates on McGaw's use of the word *gender*. Initially the feminist approach focused on *women* and technology; however, because women's relationship to technology differed from that of men by virtue of their sex and because a set of socially constructed proscriptions was at the heart of the difference, the term *gender* emerged as a more useful conceptual category. The concept of gender, Rothschild believes, allows us not only to analyze such proscriptions but also to probe into technology more deeply. This analysis can occur in any of six ways: by introducing issues of race and class into technology studies, by extending research cross-culturally, by enlarging the subject matter to include neglected traditional technologies, by modifying defective theories of technological development, by challenging outdated concepts of technological change, and by raising new concerns for the philosophy of technology. Rothschild concludes by reminding us that "gender perspectives by their very nature are multi-disciplinary and interdisciplinary." Consequently, such perspectives contribute to the attainment of an aim that has begun to assume considerable value among historians of technology.

Part of the appeal of a more interdisciplinary perspective lies in the intrinsic dynamics of the emergence of new academic specialization. Because their subject matter is often perceived as too obscure for survey courses, such emerging specialties often find themselves removed from mainstream undergraduate education. There is also a perception that as scholarly specialties develop they become increasingly inaccessible to the uninitiated. Yet, in the case of the history of technology, Darwin Stapleton finds that in general this perception does not hold true. Because the field has no "standard" textbooks, teachers frequently use articles in journals, particularly *T&C* and also, because of the inherently visual nature of the much of technology, well-illustrated magazines such as *Scientific American*. Curricular concerns, which have increasingly occupied SHOT members, began to become formalized in 1977–78 with the creation of the Technology Studies and Education (TS&E) interest group. TS&E has subsequently compiled two editions of model course syllabi, *The Machine in the University*.[6] The success of this enterprise testifies to the widespread interest in matters of curriculum, as does the popularity of such curriculum-oriented newsletters as the *Science, Technology and Society Curriculum Development Newsletter, The Weaver,* and *History of Science: News and Views*.

In addition to those scholars whose aim is to foster perspectives on technology that are more interdisciplinary, this volume also includes an explicit plea for a more international stance. In arguing for "a perspective less nationalistic in outlook and tone," David Hounshell suggests that a "consensus" historiography has continued to prevail among historians of technology in the U.S. far longer than among other specialists. The emphasis on the London Crystal Palace Exhibition of 1851 as a "watershed" and on the rise of the "American system of manufactures" emerged as part of an interpretation that posited an emphasis on American uniqueness and superiority. With only a few notable exceptions, such notions have persisted even among historians sensitive to contextualism. Hounshell calls for the cultivation of a perspective that transcends national, and especially American, boundaries and for a comparative history that takes much fuller account of European contributions to a broad range of developments, both technical and theoretical. Only then, he concludes, will we achieve "a richer, more textured history of American technology."

Like Hounshell, Brooke Hindle is also concerned with our breadth of understanding—in his case, as it relates to the function of history more generally. But, if Hounshell is to a large extent arguing against the "consensus" approach that he sees persisting in the historiography of technology, Hindle's precepts are different. History of technology's rapid establishment as a significant specialty is owing in part to a widespread recognition of technology's central role in contemporary society. For that very reason, Hindle argues, the history of technology must be "increasingly integrated into our understanding of human history as a whole." Because the history of technology is not well understood even by other historians, that integration must be performed by historians of technology. To do this properly, historians of technology must remain sensitive to general historical currents. However, certain historians interested in technology who *have* had an impact on general history are scholars of the sort Hindle refers to as "darksiders," adherents of a negativistic perspective whose emergence he links to the antiestablishment fervor of the late 1960s and early 1970s and to the rise of the New Left. Hindle does not deny the importance of critical thinking or of exploring the negative aspects of technology as well as the positive ones. But, inasmuch as history is society's memory, in order for society to function well (as for an individual to function well), that memory must be primarily positive. A memory that concentrates on defeats and failures is "a route to disaster." Hindle, then, not only believes that the history of technology must be integrated into our general concept of history, but also that, in the interest of a society that functions in a healthy manner, that recorded history should remain essentially positive.

Hindle's conclusion is controversial; indeed, Hounshell and some of the other authors in this volume would take issue with him. Yet, we believe that such a situation is indicative of a healthy state of affairs for the history of technology. Mel Kranzberg, who as editor of *T&C* liked nothing better than constructive controversy, would certainly agree. This collection is somewhat different from the usual Festschrift in that, in addition to its honorific function, it was conceived as a reflection of a discipline—its past, its current state, and its potential. About where it has come from and where it is, there can be some degree of consensus, but about where it ought to be headed, there is bound to be considerable conflict.

Above all, this volume was conceived in a spirit of self-examination. That being the case, we have decided it would be eminently fitting to include the thoughts of the man to whom the book is dedicated, Melvin Kranzberg. Hence, the final essay is Mel Kranzberg's 1985 SHOT presidential address, "Technology and History: 'Kranzberg's Laws'." Many of us may feel that Mel conveyed countless messages over the years, but, by his own lights, every one of them was rooted in one primary theme: that of "the significance in human affairs of the history of technology and the value of the contextual approach in understanding technical developments." Ultimately, then, even though this book was written *for* Mel Kranzberg, it is Mel himself who has said it best. For it is this primary theme that holds together not only this book but, indeed, the whole field in whose creation Mel was so instrumental.

NOTES

1. Obviously exceptions existed; Lewis Mumford, Abbott Payson Usher, Sigfried Giedion, and a few others were well known and well regarded.

2. For a brief description of these early events, see John Staudenmaier, *Technology's Storytellers: Reweaving the Human Fabric* (Cambridge, Mass.: MIT Press, 1985), pp. 1–3. This book is an interpretive study of the emerging intellectual character of the history of technology through a detailed analysis of the articles published in the first twenty years of *Technology and Culture*.

3. For good summary listing of such essays, see Carroll W. Pursell, Jr., "History of Technology," in *A Guide to the History of Philosophy and Sociology of Science, Technology and Medicine*, 2d ed., ed. Paul Durbin (New York: Free Press, 1984), pp. 103–4; Staudenmaier, *Technology's Storytellers*, pp. 7–8, 13–14; and Merritt Roe Smith's essay in this volume, esp. n. 2.

4. Edwin T. Layton, Jr., "Mirror Image Twins: The Communities of Science and Technology," *Technology and Culture* 12 (October 1971): 562–80.

5. Thomas P. Hughes, *Networks of Power: Electrification in Western Society, 1880–1930* (Baltimore: Johns Hopkins University Press, 1983); David Hounshell, *From the American System to Mass Production, 1800–1932* (Baltimore: Johns Hopkins University Press, 1984); David F. Noble, *Forces of Production: A Social History of Industrial Automation* (New York: Alfred A. Knopf, 1984); Ruth Schwartz Cowan, *More Work for Mother: The Ironies of Household Technology from the Open Hearth to the Microwave* (New York: Basic Books, 1983).

6. Terry S. Reynolds, ed., *The Machine in the University*, 2d ed. (Bethlehem, Pa: SHOT and Science, Technology and Society Program, Lehigh University, 1987).

Melvin Kranzberg: An Appreciation

CARROLL W. PURSELL, JR.

Any introduction of Mel Kranzberg, the consummate after-dinner speaker and raconteur, can appropriately begin with that disclaimer he must have heard on many occasions—that he, indeed, needs no introduction. Not only here in the United States but also in Europe and Asia, he more than any other person represents the field of the history of technology. By his scholarship; his frequent speaking engagements; his founding of institutions; his service on innumerable scholarly, educational, and governmental bodies; and, most important, by the very energy, enthusiasm, and generosity of his character, he has done more than any other person to nurture and shape the field here and abroad.

In the United States, two major components of the field—the Society for the History of Technology and its journal, *Technology and Culture*—are of his making. In January 1958 Kranzberg was chair of the committee that gave birth to the society, and he was its secretary until 1975. The first issue of *Technology and Culture* appeared in the winter of 1959, and his editorship of that distinguished journal lasted from its inception until 1981.[1] In 1985 he looked back and confessed that he had "been conveying basically the same message for over thirty years, namely, the significance in human affairs of the history of technology and the value of the contextual approach in understanding technical developments."[2]

On the international scene, Kranzberg has been a vice president of the International Committee for the History of Technology since its organization in Paris in 1968. Along with his Soviet counterpart, he has provided much of the continuity and energy that has made that body a growing influence on the worldwide interest in the history of technology.

Kranzberg has never been content, however, merely to preach to and serve the already converted. A primary audience has always been those who think technology is either good or bad, or, worse yet, those who do not give it any thought at all, and it is to these that he directs his true missionary's zeal. Serving on national boards, testifying before congressional committees, granting interviews with the press, buttonholing those with authority and the power to effect change, and making uncountable speeches, Mel Kranzberg has spent his career as what we would now call a public historian. His message has been consistent: history is not a subject of merely academic interest. Those who hope to understand and shape an age and a people heavily under the influence of technological change must first study that change through time.

But it is the spirit rather than the letter of the law that giveth life, and it is Kranzberg the gentleman as much as Kranzberg the scholar who has left an indelible mark on the history of technology. Everyone who gave a paper at a SHOT meeting while he was secretary soon after received a letter from him asking for the name and address of the appropriate dean at the speaker's home campus. Mel took the time to write to each, praising the scholarly contribution and congratulating the dean on his or her fine faculty. His endless and effective letters of recommendation and general praise were, he has always maintained, not due to his own generosity and magnanimity, but merely the result of having a secretary who had to be kept busy at something. We all know better.

While still in the floodtide of a career serving the history of technology, Kranzberg has already won every honor available—he has been secretary, editor, president, and recipient of the da Vinci medal, the society's highest award, given annually to an individual who has made an outstanding contribution to the history of technology through research, teaching, publication, or other activities. Although another generation has stepped forward to carry on the duties he invented and performed so gracefully and successfully for years, new secretaries and editors find that the mail still comes addressed to him, and this is fitting. Throughout the world of scholarship, the name Mel Kranzberg is synonymous with the history of technology.

NOTES

1. See Melvin Kranzberg, "At the Start," *Technology and Culture* 1 (Winter 1959): 1–10, and "Passing the Baton," *Technology and Culture* 22 (October 1981): 695–99.
2. Melvin Kranzberg, "Technology and History: 'Kranzberg's Laws,'" *Technology and Culture* 27 (July 1986): 544.

Contributors

PAUL T. DURBIN is a professor in the philosophy department and the Center for Science and Culture at the University of Delaware, general editor of *A Guide to the Culture of Science, Technology, and Medicine* (1980, paper 1984), editor of *Research in Philosophy & Technology* (vols. 1–8 and supplement, 1978–1985), and coeditor (with Friedrich Rapp) of *Philosophy and Technology* (1983). He has published numerous articles on the ethical and social responsibilities of engineers and scientists.

EUGENE S. FERGUSON is a former engineer and graduate of Carnegie Tech. He taught at Iowa State University and at the University of Delaware in the history of technology program. He also was a curator at the Smithsonian Institution and at the Hagley Museum.

BROOKE HINDLE is historian emeritus at the National Museum of American History of the Smithsonian Insitution where he served earlier as director. He taught previously at New York University and served for a time as dean. His history publications are primarily on early American technology and science.

DAVID A. HOUNSHELL is professor of history at the University of Delaware. He is the author of *From the American System to Mass Production, 1800–1932: The Development of Manufacturing Technology in the United States* (1984), which received the 1987 Dexter Prize of the Society for the History of Technology, and, with John Kenly Smith, Jr., of *Science and Corporate Strategy: DuPont R & D, 1902–1980* (1988). For 1987–88, he was a Marvin Bower Fellow at the Harvard Business School.

THOMAS P. HUGHES is Mellon Professor of the history and sociology of science at the University of Pennsylvania and Torsten Althin Professor of the History of Technology and Society at the Royal Institute of Technology, Stockholm. His most recent book, *Networks of Power* (1983), is about the electrification of the western world and won the Society for the History of Technology's Dexter Prize in 1985. He is now writing a book about technology and culture, 1880–1980.

EDWIN T. LAYTON, JR., is a past-president of the Society for the History of Technology. His education was in history at UCLA, and his early interest lay in engineering professionalism, ethics, and sense of social responsibility. His book *The Revolt of the Engineers, Social Responsibility and the American Engineering Profession* won the Dexter Prize of the Society for the History of Technology in 1971. Since that time he has been studying the interaction of science and technology and the nature of technological knowledge. He is currently working on the development of the hydraulic turbine in America as a case study of this interaction.

JUDITH A. MCGAW is associate professor of the history of technology in the Department of History and Sociology of Science, University of Pennsylvania. Her publications include, *Most Wonderful Machine: Mechanization and Social Change in Berkshire Paper Making, 1801–1885* (1987) and several articles on women and technological change in history. She is currently studying agricultural and domestic production in southeastern Pennsylvania, 1750–1850.

ARTHUR P. MOLELLA is chairman of the Department of the History of Science and Technology at the National Museum of American History, Smithsonian Institution, and former book review editor of *Technology and Culture*. He is pursuing research on the relationship of philosophy and conceptualizations in science and technology.

ROBERT P. MULTHAUF is a former director of the National Museum of History and Technology, former editor of *Isis*, and former Senior Historian at the National Museum of American History, Smithsonian Institution.

LIZ PALEY is completing a Ph.D. in the history of technology at Case Western Reserve University; her research specialty is technology and the press. She has also worked on a history of Cleveland's University Circle, contributed to the *Encyclopedia of Cleveland History,* and currently teaches at the Cleveland Institute of Art.

CARROLL W. PURSELL, JR., is professor of history and director of the Program in the History of Technology and Science at Case Western Reserve University. He received his M.A. from the University of Delaware (Hagley Fellow, 1958) and his doctorate from the University of California, Berkeley. Secretary of the Society for the History of Technology from 1975 to 1984, he has taught at Lehigh University and the University of California, Santa Barbara. Currently he serves as vice-president of ICOHTEC.

STEVEN C. REBER is a graduate student in the Program in Science, Technology and Society at MIT. He holds a B.S. degree in electrical engineering from MIT and an M.S. in applied physics from Columbia University. He is currently engaged in research for his doctoral dissertation, a study of the ceramics industries of Britain and France during the eighteenth century.

JOAN ROTHSCHILD, professor of political science and former coordinator of Women's Studies at the University of Lowell, has been a member of the university's Technology, Society and Values program since its inception in 1977. A contributor to numerous journals and volumes in both women's studies and the technology fields, she is the author-editor of *Women, Technology and Innovation* (1982) and *Machina Ex Dea: Feminist Perspectives on Technology* (1983), and author of *Teaching Technology from a Feminist Perspective* (1988). Her current research interest is in exploring the impact of the human perfection ideology on the new reproductive technologies.

BRUCE SINCLAIR is president of the Society for the History of Technology and Kranzberg Professor of the History of Technology at the Georgia Institute of Technology. He is the author of *A Centennial History of the American Society of Mechanical Engineers, 1880–1980* (1980), and *Philadelphia's Philosopher Mechanics: A History of the Franklin Institute* (1974), which won the Society for the History of Technology's Dexter Prize in 1975.

MERRITT ROE SMITH is professor of the history of technology at MIT. He is the author of *Harpers Ferry Armory and the New Technology* (1977) and the editor of *Military Enterprise and Technological Change: Perspectives on the American Experience* (1985).

DARWIN H. STAPLETON is a graduate of Swarthmore College, and earned his Ph.D. in 1975 at the University of Delaware while holding a Hagley Fellowship. From 1976 to 1986 he was at Case Western Reserve University teaching the history of science and technology, and American studies. Concurrently he was an editor of The Papers of Benjamin Henry Latrobe. Presently he is the director of the Rockefeller Archive Center and an adjunct professor at Rockefeller University.

JOHN M. STAUDENMAIER, S.J., is associate professor of history of technology at the University of Detroit. His book, *Technology's Storytellers: Reweaving the Human Fabric* (1985), won the 1986 Alpha Sigma Nu National Jesuit Book Award in the Sciences. He is currently working on an interpretation of United States technological style in the twentieth century.

In Context

History and the History of Technology

Through the Looking Glass; or, News from Lake Mirror Image

EDWIN T. LAYTON, JR.

My theme is the nature of technological knowledge. However, I would like to begin with a parable based on Lewis Carroll's *Through the Looking Glass.* In it Alice discovered another reality beyond the looking glass, one where things were different in some rather interesting ways. I wish to report on a rare variant of Carroll's classic published by an obscure technological press. In the edition I used, the rooms on both sides of the looking glass were filled with people at work: one side was labeled "science" and the other side "technology."

In this version Alice had difficulty in telling one side from the other. The people seemed to be doing the same sorts of things: they fiddled with computers, manipulated various pieces of apparatus, made drawings and calculations, and argued with one another. On both sides some of the people performed experiments and developed theories while others designed technological devices. Although the proportion of people engaged in these tasks differed somewhat, Alice found it very confusing, especially considering that the initial output in both cases appeared to be information.

Not only were the goings-on similar, but Alice also found that now she had to run as fast as possible just to stay in one place on both sides of the looking glass. In addition, the Jabberwocky was present in both worlds, and the people in both seemed about equally eager to supply him with still sharper and more deadly teeth and claws. There were differences, but they were often rather subtle. When Alice held up the book titled *Thermodynamics,* it changed to *Engineering Thermodynamics* in the looking glass. Alice became a bit confused and asked whether it really made any important difference what side of the looking glass people were on.

At this point in the variant edition, an alien from another galaxy appeared

Edwin Layton originally presented this paper as his presidential address for the Society for the History of Technology on 24 October 1986 in Pittsburgh, Pennsylvania. It was subsequently published in *Technology and Culture* 29 (July 1987): 594–607, and is reprinted here by permission.

with a time machine. (Alice thought his name was "Kranzberg," but she must not have heard it correctly. There cannot be two of them, she thought.) When they traveled in the time machine, the images in the mirror became different. The further they traveled backward in time, the more unlike the images became until Alice could see little similarity between the two.

On the science side the big particle-accelerators and giant telescopes disappeared, as did nearly all of the scientists engaged in industrial research and design. Instead, Alice saw a Cambridge University professor examining light with a prism and an Italian professor experimenting with balls of different weight; further back there was a long line of Greek philosophers, Babylonian priests, and so forth. They seemed to be preoccupied with building theories, or simply with "knowing."

On the technology side the changes were even more dramatic. The theories and experimentation diminished, but they never quite disappeared completely. The engineers shrank from a multitude almost filling the room to a mere handful. After a long time they too faded away. In their place was an even longer line of hunters, gatherers, farmers, artisans, and artists. They appeared to be working people making or doing things.

But while the differences grew greater with time, Alice noticed that at certain times the same people tended to appear on both sides of the looking glass. There were, for example, a number of eighteenth-century philosophers: she thought she recognized James Watt and Benjamin Franklin. Also present on both sides were a number of Renaissance artist-engineers including Leonardo da Vinci. At an earlier period there were cathedral-building masons and, still further in the past, Hellinistic engineers.

The probability of someone being on both sides of the looking glass appeared to increase as she approached the present. This was particularly the case in industrial research laboratories. In fact, the mirror became so permeable and difficult to see that many people denied that it existed. Alice concluded that the difference was sometimes important and sometimes not, depending on the questions asked. She noticed that some people on both sides seemed to be obsessed with determining who got credit for what. She concluded that this was not, by itself, a very important question.

As in the original version, Alice passed through the looking glass. She found that what could be seen from the old room was the same. But the parts she could not see from the old room were quite different in unexpected and delightful ways. Workers, farmers, artists—all kinds of people of both sexes—were still there. They were just as numerous and as hard working as they had been in the distant past. But when viewed from the science side of the present time, they had been pushed down out of sight by the weight of the giants standing on their shoulders.

From the science side it had appeared to be a masculine affair, no matter which way she looked. However, from the technology side of the mirror she saw that both sexes were well represented in a variety of roles.

Alice found that the perspective provided by technology was especially suited to studies of gender. Indeed, Alice saw that the study of gender was one part of a very rich body of social history that is particularly accessible from the technology side of the mirror. Some were community studies deriving partial inspiration from sociology and anthropology. Others used the systems approach to understand technology and its impact on society. Alice saw that the perspective of technology provided a great many exciting and interesting ways of looking at history. Most were not motivated by a desire to understand techology as knowledge or the role of science in technology, but she saw that such studies often benefited by understanding the knowledge dimensions of technology.

Now, in this version of *Through the Looking Glass,* Alice decided that the technology side was much more fun, more dynamic, more open to new ideas, to novel methodologies, to women and to younger people. So she stayed on the technology side of the looking glass and founded an organization with the acronym WITH, and lived happily ever after. (As noted, this edition was produced by a technological press.)

In one volume there was a historical note written in years later. In it, Alice was vexed to discover that some very good people had misunderstood her. She finally concluded that it was her own fault for being too cute in using metaphors, and not being clear enough about her assumptions.[1] But she also noticed that these people often did outstanding work without her help. She concluded that the progress of scholarship is collaborative and does not depend on any one person. It was a much needed lesson in humility for Alice.

That is the parable from Lake Mirror Image, where all the engineers are strong (except perhaps when it comes to blowing whistles), and all the scientists are good looking (if not always looking for the Good). And all the historical publications are above average (particularly if they appear in *Technology and Culture*).

*   *   *

In case you had not noticed, this parable is intended to introduce the topic of the nature of technological knowledge, a theme that is, in turn, closely bound with the interaction of science and technology.

Although studies on the relation of science and technology have appeared in the literature of history of technology from the outset, there has been a renewed interest in the topic, with considerable progress made in the last twenty-five years. The debate now includes related disciplines, the sociology and philosophy of science and technology. A certain degree of consensus on a new model of science-technology relations has emerged that is reflected in different ways in three recent surveys by Barry Barnes, John Staudenmaier, and Alexander Keller and in several articles in a work edited by Paul T. Durbin.[2]

This interactive model, as it is sometimes called, has gained widespread support. One of the clearest and most concise statements of it is provided by Barnes, a British sociologist of science.[3] Barnes playfully refers to two models, one the "bad old days," the other our present understanding. In the "bad old days" technology was thought to be hierarchically subordinate to science. This model assumed that technology derived its intellectual substance from science, most particularly from the discoveries of the basic sciences, and that without this intellectual sustenance the advance of technology would cease. The role of science was to make discoveries and create knowledge, while the role of technology was to apply these discoveries.

In contrast to the "bad old days," Barnes sees a new interactive model dominating the present. Technology, no longer considered subordinate, is recognized as an autonomous, coequal community. The relationship between science and technology is symbiotic, egalitarian, and interactive. Technology and science both make use of the products of the other from time to time, in the mutually beneficial way characteristic of symbiotic relationships. Technology can draw on existing technology as well as existing science. Technology's form of cognition is now seen as creative and constructive, like that of science. Characteristic of the present view, science is no longer seen as making and evaluating discoveries in a manner that is substantially independent of context. Instead, both science and technology should be evaluated in terms of the ends pursued in the appropriate social context.[4]

More recently, John Staudenmaier has come to somewhat similar conclusions, though placed in a different framework.[5] For him as for Barnes, the most significant result of the debate has been to help establish another new consensus showing that science was not, as is sometimes claimed, the only objectively valid form of knowledge.[6] Barnes had explicitly proposed extending the view of science implied by this new model to the relations of science with other branches of society.[7] He credits the creation of the interactive model to historians of technology, including Donald Cardwell, Thomas P. Hughes, and the late Derek J. de Solla Price.[8]

John Staudenmaier, however, has suggested an ironic twist. The science-technology interaction appeared to be critical in the early days of the history of technology precisely because of the absolutist claims made by the more extreme apologists for science. Historians of technology had to demolish the myth that technology was no more than applied science in order to establish the legitimacy and autonomy of their own discipline. But once they were successful in this, they found that it was not really the interaction with which they were primarily interested. Rather, the emphasis lay in the nature of technological knowledge.[9] I agree with Staudenmaier, although I have a few small qualifications.

The issue can be clarified by distinguishing between the three types of discussion of this interaction. First, many studies noted that, in the field studied, practice led theory rather than the reverse; this category includes virtually all the founders of the modern, critical discipline of history of technology. Second, various historical studies viewed the science-technology relationship as the essential starting point for an understanding of the nature of technological knowledge; in these studies the interaction of science and technology is not necessarily a significant concern. Third, historians of technology have offered rebuttals of statements of the applied-science model of technology, sometimes illustrated by specific historical case studies.

For the history of technology as a discipline, the first two categories of study were, and remain, the more important. The primary concern of historians of technology is their own discipline. But from a broader social perspective, as Barnes points out, the role of historians of technology has been important in developing a new model for understanding the cultural role of science and technology, which is having an influence far beyond the confines of our discipline. In this latter category, the solid case studies have been the most influential.[10]

These types of history of technology cannot always be separated. The term *symbiosis* was intended to be taken literally. That is, the interaction is an essential, not an accidental, part of both science and technology, which accounts for many fundamental characteristics of both. In it, I believe, lies the explanation of the dynamic growth of both science and technology since the Renaissance, and perhaps even earlier. I said my qualifications were small. I do not wish to claim that all studies in history of technology should focus upon the interaction with science. Quite the contrary, only a minority need start there. My caveat is that there is a significant subset in history of technology where an understanding of the role of science is indispensable for writing technological history.

Many topics in recent engineering history simply cannot be understood except in the context of the symbiotic interaction of science and technology. Ronald Kline's article about the development of the induction motor is an example. He is not primarily concerned with ideological questions. His data, as distinct from his conclusions, could be used by proponents of various models. He shows that laborious research by engineers was needed to gain the understanding needed to design alternating-current motors.[11]

This engineering work clearly rested on foundations provided by research in the basic sciences by such figures as Arago, Faraday, and Maxwell. In the end, engineers produced their own theories (notably that of Charles P. Steinmetz), which in a certain sense broke with Maxwell and the basic science heritage. As a contemporary engineer, Arthur E. Kennelly, commented:

> I, for one, am opposed to other than engineering methods for treating engineering problems. I believe that the method which has just been discussed by Mr. Steinmetz before us on the blackboard is far more capable of giving us the engineering properties of induction motors than are the methods based on Maxwell's equations.[12]

This statement did not indicate that Maxwell's work was incorrect, or that it had no impact on technology. Quite the contrary was the case. But it did indicate the maturity of engineering theory in this particular area. Scientists favor idealized theories for maximum understanding; engineers must build a bridge between pure theory and concrete, complex artifacts or systems working in the real world. These intermediate structures of cognition are design and the sciences that serve design. In the case of engineering, these can be variously termed *engineering sciences* or *engineering theories*. Engineers often prefer the latter, as in the theories of circuits, elasticity, structures, combustion, stability, control, plasticity, and so forth.

But while differing from the basic sciences in a number of important characteristics, engineering sciences or theories seem to share the same general form. They are analogous ways of organizing knowledge. Indeed, engineers have borrowed heavily from the examples provided by the basic sciences in constructing their theories.

Another indication of the arrival of a new understanding of the relations of science and technology is the appearance of Alexander Keller's survey, "Has Science Created Technology?"[13] To this can be added a broad survey of the history, sociology, and philosophy of the cultures of science and technology and medicine edited by Paul Durbin. Treating science and technology as subcultures is, of course, the essense of the interactive model. Of particular relevance are the essays contributed by Carroll Pursell, Carl Mitcham, and Jerry Gaston.[14] Keller and the authors in the Durbin volume take no explicit stand, but the aggregate effect is, I believe, to provide strong support for the interactive model.

Two different sets of claims regarding the impact of science on technology have been proposed. The strong claim holds that science created technology, whereas the weaker claim (associated with R. A. Buchanan and repeated by Keller, but in existence since the time of Bacon and the Royal Society) holds that science "will lay a succession of golden eggs, and that society should pay to understand how nature works in order to exploit the potentialities of nature."[15] The stronger claim has, I believe, been refuted by recent scholarship. The weaker claim has, however, been sustained.

Not only have the basic sciences produced a series of "golden eggs," but the interactions between the basic sciences and technology have also been both more complex, more subtle, and more fruitful than previously suspected. The intereactive model has shown that, besides the direct influences (which exist and are important, even if they have sometimes been overstated), the indirect impact of science on technology has been very important. It would, of course, be equally true to point out that technologists over the centuries have produced "golden eggs" of great value to mathematics, physics, chemistry, and biology. Here too indirect interactions have been very important. The sciences as well as technology are best seen as the products of their mutual coevolution, at least in the modern era. That is, the two really are symbiotic with one another. Neither is the exclusive property of any single community.

What then is the nature of technological knowledge, and how does it differ from that in the basic sciences? A. R. Hall has argued that technological knowledge is knowledge of how to do or make things, whereas the basic sciences have a more general form of knowing.[16] When he considers what science has contributed to technology he lists four items: mathematical analysis, the method of establishing facts by controlled experiment, the knowledge of relevant natural laws such as thermodynamics, and acquaintance with new natural phenomena such as electromagnetism or catalysis. As Keller comments: "This is not too different from Dr. Layton's view of the matter."[17] I believe this is a reasonable inference. There is no fundamental disagreement on these issues, although some differences in detail or of interpretation are significant and worth clarification.

I have already discussed the two-way traffic in ideas and methods between the communities of science and technology. I have noted the fact that engineers and other technologists have developed their own theories or sciences in the service of design and the fact that these are structured in a manner analogous to the basic sciences. However, I believe that all such knowledge is in an ultimate sense a part of "knowing how."

This distinction between knowledge in the abstract and "knowing how" is fundamental. Technological knowledge can be divided into at least three compartments: technological science (or theory), design, and technique. All of these come into focus in design, the critical act of synthesis in technology. As noted in a handbook on design, "Design is the essential purpose of engineering."[18] The ability to design is the most critical requirement for membership in the professional grades of most American professional engineering societies (where the state or private organizations do not provide a ready-made differentiation between professional engineers and nonprofessional technical workers).[19]

The reason technologists value design so highly is not difficult to find. A design embodies the knowledge needed to produce a technological device or a system. It constitutes the cognitive bridge across a spectrum from abstract, idealized conceptions to the concrete, highly complex products of technology existing in the real world. It is "knowing how" at the highest level. In the American vernacular, "know-how" usually refers to knowledge at the design level. Of course, as Frederick Taylor and Henry Ford demonstrated, systems of production have become as much the products of conscious design as are the final products. Product and process innovations both involve design.

Technological knowledge can exist at all levels. Engineering theorists and designers need to develop and use theoretical knowledge. In the modern period, this theory is increasingly scientific. Technological knowledge, however, can extend along a hierarchy from the most abstract to the most concrete, from the most intellectually demanding to the least demanding routine jobs. We would not expect the clerk who uses a word processing computer to have the same understanding of the theory of computers as that possessed by the designer of the computer. Nor would the theory employed by this designer be identical to that of the mathematician or physicist interested in computers.

Two points need to be made here. Working people are intelligent and make important contributions, particularly in technique and skill. Technology must involve the mental efforts of many people at all levels; brainwork is not all concentrated in a tiny elite. Nor is the dignity or worth of the individual dependent on their location within a hierarchy.

Second, attempts to compare two communities or disciplines by contrasting the most intellectually advanced practitioners of the one with the least intellectually advanced of the other are invalid. It would be fruitful and illuminating to compare the knowledge and style of a great computer designer such as Seymour Cray, the inventor of the supercomputer, with the more abstract concerns of a mathematician such as Alan Turing. However, comparing Einstein's knowledge of science with a child's knowledge of how to ride a bicycle does not advance our understanding.

It has proved very fruitful to contrast theory and experiment in technology and engineering with parallel activities in the basic sciences. The differences are subtle, but real and rewarding. Perhaps no one has done more to explicate the special modalities of the scientific method in engineering design than Walter Vincenti. His classic studies of the method of parameter variation in the design of the Britannia Bridge and in Durand's propeller tests, as well as in other cases, and his study of the engineering use of the concept of control volume have given precise examples of engineering theory and experiment in the service of design.[20] Studies of this sort do not prove that one community's approach is superior to that of another; they suggest only that form follows function. Technological knowledge is tailored to serve the needs of design. Knowledge in the basic sciences is shaped by the desire to construct the most general and comprehensive theories.

This point can be illuminated by a recent paper in chemical engineering by Octave Levenspiel. Theory in this field, like electronics, has become very sophisticated and mathematical. It is not uncommon now to find chemical engineers using not only the usual differential equations but more advanced methods as well. Given the requirements of this sort of engineering theory, I doubt that a meaningful distinction can be drawn between chemists and engineering theorists in terms of motivation. All theorists appear to take delight in problem solving. A superficial observer might conclude that what theoretically inclined chemical engineers do is science, not technology. And yet chemical engineering continues to be a field of technology, for all its mathematical trappings. One of the most critical forces driving chemical engineering to this more scientific and theoretical approach was the changing needs of chemical technology. In particular one important beginning can be traced to the need for a better understanding of catalysis and associated transport processes in the catalytic cracking of petroleum. In general, catalysis is another area where practice generally led theory, at least until very recently.[21]

Practitioners in each community can be expected to value their own approaches over those of others. This merely reflects the value systems that distinguish the communities in the first place. Although this point may appear obvious, it is worth further clarification.

Levenspiel, noting that chemical engineers and chemists both were concerned with kinetics of chemical reactions, asked what the chemical engineer had to contribute. His answer was "plenty; in essence a clear understanding of how to use this information to tell how a reactor will behave."[22] Later in the same paper, in praising the developing merger of chemcial engineering and biotechnology, Levenspiel insisted that the special language developed by chemical engineers should become dominant in the new technology. To make his point he contrasted the classical scientific equation to describe the growth of microorganisms with the formulation natural to chemical engineers. In his words: "Why this emphasis on technical language? Because the language we think in profoundly affects the thoughts we think. . . . Likewise Chemical Reaction Engineering is the language of reactor design."[23]

That is, given the goals of the technologist, and the social context, then the values are virtually inescapable. These goals and values, along with the state of the art, have a profound influence on the form and substance of technological theories and methods. There is no reason to say that one community or one mode of thinking is somehow inferior or superior to another. Such judgments, when they occur, usually reflect attempts to apply the values of one community to the work of a different community.

At various times scientists and engineers have occasionally made disparaging remarks about one another. These comments are important and interesting in indicating the values of the two communities, their dynamics, and their interactions.[24] They do not, however, indicate that science and technology are at war with one another, but rather that symbiotic relations, like personal ones, may engender some friction.

The interactive model of science-technology relations is not identitical with the symbiotic model. The latter implies a stronger interaction than some other interpretations do. Symbiotic versions of the interactive model imply coevolution. To adopt such a position does not mean a return to the "bad old days" (in Barnes's terms). Nor does it mean that all studies in history of technology should start with the science-technology interaction. Only certain levels of discourse are affected. Coevolution deals with the large-scale structure of the development of both science and technology. We already have evidence for such coevolution in Donald Cardwell's works. In this view, thermodynamics is the joint product of the communities of technology and of science: it cannot be fully understood by reference to science alone or to technology alone.[25] There are many exciting possibilities for further research on this and related topics. The purpose of the interactive model then was not simply to refute the "applied science" theory. Rather, the interactive model removes encrustations of myth and ideology, while opening the way for further studies on the subject.

NOTES

1. My work on the interaction of science grew directly out of the research for *The Revolt of the Engineers, Social Responsibility and the American Engineering Profession* (Cleveland: Press of Case Western Reserve University, 1971; Baltimore: Johns Hopkins University Press, 1986). One of my most fundamental assumptions was the theory of orientation (e.g., with respect to the potentially conflicting demands of professions and employers) developed by a number of sociologists and applied in a book I assumed was well known, William Kornhauser, *Scientists in Industry* (Berkeley and Los Angeles: University of California Press, 1962). I should have cited this work in my early papers on the interaction of science and technology. Kornhauser explicitly applies the theory to scientists and engineers employed in industrial bureaucracies. I started from a basic paradox, that as science and technology came closer together and interacted more intimately, the boundaries separating them (as defined by professional societies) became more sharply defined. Unfortunately, I did not tell my readers this. Keller (cited below) has made some telling points against the excessive reliance upon metaphor in discussions of the interaction of science and technology.

2. Barry Barnes, "The Science-Technology Relationship: A Model and a Query," *Social Studies of Science* 12 (February 1982): 166–71; John M. Staudenmaier, S.J., *Technology's Storytellers: Reweaving the Human Fabric* (Cambridge, Mass.: MIT Press, 1985); Alexander Keller, "Has Science Created Technology?" *Minerva* 22 (1984): 160–82; and Paul T. Durbin, ed., *A Guide to the Culture of Science, Technology, and Medicine* (New York: The Free Press, 1980).

3. Barnes, "The Science-Technology Relationship," pp. 166–68.

4. Ibid., p. 168.

5. Staudenmaier, *Storytellers.*

6. Ibid., pp. xvii–xviii, 169–70, passim; John M. Staudenmaier, "What SHOT Hath Wrought and What SHOT Hath Not: Reflections on Twenty-five Years of the History of Technology," *Technology and Culture* 25 (October 1984): 709–12. See also Michael Mulkay, "Knowledge and Utility: Implications for the Sociology of Knowledge," *Social Studies of Science* 9 (1979): 63–80.

7. Barnes, "The Science-Technology Relationship," pp. 168–71.

8. Ibid., p. 171. To these a great many other names could be added, too many to enumerate here. Speaking subjectively, I can say that I was influenced and assisted by a number of the founders of our discipline in addition to Cardwell, Hughes, and Price, including John G. Burke, Carl W. Condit, Eugene Ferguson, John R. Harris, Louis C. Hunter, Melvin Kranzberg, Ladislao Reti, Hunter Rouse, Robert P. Multhauf, John B. Rae, Cyril S. Smith, Abbot Payson Usher, and Lynn White. I drew direct inspiration from many contemporaries (not all of whom agreed with me), including Hugh G. J. Aitken, Edward W. Constant, Vernard Foley, Charles C. Gillispie, Barton C. Hacker, Brooke Hindle, Reese Jenkins, Alexander G. Keller, Otto Mayr, Robert C. Post, Friedrich Rapp, Leonard S. Reich, Herbert Simon, Merritt Roe Smith, Norman A. F. Smith, and George Wise, to mention only a few. But I have been closest to and

particularly affected by the works of certain friends, some of whom were once students of mine either directly or indirectly, and who were, in turn, my teachers, consciously or unconsciously. This group includes David P. Billington, James Brittain, David F. Channell, Eda Fowlks Kranakis, Robert Mark, Terry S. Reynolds, Bruce Sinclair, Robert E. Schofield, John H. Weiss, Peter Weingart, and last but far from least, Walter Vincenti.

It is impossible within the present scope to list all of the relevant works. For brevity I have restricted myself by citing only one work by each scholar noted, though, in fact, I was often more influenced by the cumulative effect of many works than by any single item. Here is my subjective "short list" for whatever it may be worth. First, the founders: John G. Burke, "The Complex Nature of Explanations in the Historiography of Technology," *Technology and Culture* 11 (January 1970): 22–26; Donald S. L. Cardwell, "Some Factors in the Early Development of the Concepts of Power, Work, and Energy," *British Journal for the History of Science* 3 (1966–67): 209–24; Carl W. Condit, "Sullivan's Skyscrapers as Expressions of Nineteenth Century Technology," *Technology and Culture* 1 (Winter 1959): 78–93; Eugene S. Ferguson, "On the Origin and Development of American Mechanical 'Know-How,'" *Midcontinent American Studies Journal* 3 (Fall 1962): 3–16; John R. Harris and T. C. Barker, *A Merseyside Town in the Industrial Revolution* (Liverpool: University Press, 1954); Louis C. Hunter, *Steamboats on the Western Rivers* (Cambridge, Mass.: Harvard University Press, 1949); Melvin Kranzberg and Carroll W. Pursell, Jr., eds., *Technology in Western Civilization,* 2 vols. (New York: Oxford University Press, 1967); Ladislao Reti, "Leonardo da Vinci the Technologist," in *Leonardo da Vinci, Technologist,* ed. Ladislao Reti and Bern Dibner (Norwalk, Conn.: Burndy Library, 1969); Hunter Rouse and Simon Ince, *History of Hydraulics* (New York: Dover, 1957); Robert P. Multhauf, "The Scientist and the 'Improver' of Technology," *Technology and Culture* 1 (Winter 1959): 38–47; Derek J. de Solla Price, "Is Technology Historically Independent of Science? A Study in Statistical Historiography," *Technology and Culture* 6 (Fall 1965): 553–68; John B. Rae, "The 'Know-How' Tradition in American History," *Technology and Culture* 1 (Spring 1960): 139–50; Cyril S. Smith, "Art, Technology and Science: Notes on Their Historical Interaction," *Technology and Culture* 11 (October 1970): 493–549; Abbott Payson Usher, *A History of Mechanical Inventions* (1929; rev. ed., Cambridge, Mass.: Harvard University Press, 1954); and Lynn White, jr., *Medieval Religion and Technology, Collected Essays* (Berkeley and Los Angeles: University of California Press, 1978).

For contemporaries, my subjective short list includes: Hugh G. J. Aitken, *Syntony and Spark: The Origins of Radio* (New York: John Wiley, 1976); Edward W. Constant, *The Origins of the Turbojet Revolution* (Baltimore: Johns Hopkins University Press, 1980); Vernard Foley, "Leonardo's Contributions to Theoretical Mechanics," *Scientific American* 225 (September 1986): 108–13; Charles C. Gillispie, *Lazare Carnot, Savant* (Princeton: Princeton University Press, 1971); Barton C. Hacker, "Greek Catapults and Catapult Technology: Science, Technology and War in the Ancient World," *Technology and Culture* 9 (January 1968): 34–50; Brooke Hindle, *Emulation and Invention* (New York: New York University Press, 1981); Thomas P. Hughes, *Networks of Power* (Baltimore: Johns Hopkins University Press, 1983); Reese Jenkins, *Images and Enterprise* (Baltimore: Johns Hopkins University Press, 1975); Alexander G. Keller, "The Missing Years of Jacques Besson, Inventor of Machines, Teacher of Mathematics, Distiller of Oils, and Huguenot Pastor," *Technology and Culture* 14 (January 1973): 28–39; Otto Mayr, "Yankee Practice and Engineering Theory: Charles T. Porter and the Dynamics of the High-Speed Steam Engine," *Technology and Culture* 16 (October 1975): 570–602; Robert C. Post, "In Praise of Top Fuelers," *American Heritage of Invention and Technology* 1 (Spring 1986): 58–63; Friedrich Rapp, *Analytical Philosophy of Technology,* trans. Stanley R. Carpenter and Theodore Langenbruch (Dordrecht, Holland: Reidel, 1981); Leonard S. Reich, *The Making of American Industrial Research, Science and Business at GE and Bell, 1876–1926* (Cambridge and New York: Cambridge University Press, 1985); Herbert Simon, *Sciences of the Artificial* (Cambridge, Mass.: MIT Press, 1969); Merritt Roe Smith, *Harpers Ferry Armory and the New Technology: The Challenge of Change* (Ithaca, N.Y.: Cornell University Press, 1977); Norman A. F. Smith, "The Origin of the Water Turbine and the Invention of its Name," *History of Technology* 2 (1977): 215–59; George Wise, *Willis R. Whitney, General Electric, and the Origins of U.S. Industrial Research* (New York: Columbia University Press, 1985).

The arbitrary short list of works by friends who were teachers and sometimes students includes: David P. Billington, *Robert Maillart's Bridges, the Art of Engineering* (Princeton: Princeton University Press, 1979); James E. Brittain, "The Introduction of the Loading Coil: George A. Campbell and Michael I. Pupin," *Technology and Culture* 11 (January 1970): 36–57; David F. Channell, "The Harmony of Theory and Practice: The Engineering Science of W. J. M. Rankine," *Technology and Culture* 23 (January 1982): 39–52; Eda Fowlks Kranakis, "The French Connection: Giffard's Injector and the Nature of Heat," *Technology and Culture* 23 (January 1982): 3–38; Robert Mark and Ronald S. Jonash, "Wind Loading on Gothic Structure," *Journal of the Society of Architectural Historians* 29 (October 1970): 222–30; Terry Reynolds, *Stronger than a Thousand Men, A History of the Vertical Water Wheel* (Baltimore: Johns Hopkins University Press, 1983); Bruce Sinclair, *Philadelphia's Philosopher Mechanics: A History of the Franklin Institute, 1824–1865* (Baltimore: Johns Hopkins University Press, 1974); Robert E. Schofield, *The Lunar Society of Birmingham: A Social History of Provincial Science and Industry in Eighteenth-Century England* (Oxford: Clarendon, 1963); John H. Weiss, *The Making of Technological Man, The Social Origins of French Engineering* (Cambridge, Mass.: MIT Press, 1982); Peter Weingart, "The Structure of Technological Change," in *The Nature of Technological Knowledge. Are Models of Scientific Change Relevant?* ed. Rachel Laudan (Dordrecht and Boston: Reidel, 1984); and Walter G. Vincenti, "The Air-Propeller Tests of W. F. Durand and E. P. Lesley: A Case Study in Technological Methodology," *Technology and Culture* 20 (October 1979): 712–51.

9. Staudenmaier, *Technology's Storytellers*, pp. 85, 104–5, 120.

10. Barnes, "The Science-Technology Relationship," pp. 169–71. See also Mulkay, "Knowledge and Utility," pp. 63–69.

11. Ronald R. Kline, "Scientific Electrotechnology: The Case of the Induction Motor," *Technology and Culture* 28 (April 1987): 48–62.

12. Quoted in ibid., p. 31.

13. Keller, "Has Science Created Technology?" pp. 181–82.

14. Carroll W. Pursell, Jr., "History of Technology," Carl Mitcham, "Philosophy of Technology," and Jerry Gaston, "Sociology of Science and Technology," in *A Guide to the Culture of Science, Technology, and Medicine,* ed. Durbin, pp. 70–120, 282–363, 465–526.

15. Keller, "Has Science Created Technology?" p. 160; R. A. Buchanan, "The Promethean Revolution: Science, Technology, and History," *History of Technology* 1 (1976): 73–83.

16. A. Rupert Hall, "On Knowing and Knowing How to. . . ," *History of Technology* 3 (1978): 91–103.

17. Keller, "Has Science Created Technology?" p. 178.

18. J. B. Reswick, "Foreword," in Morris Asimow, *Introduction to Design* (Englewood Cliffs, N.J.: Prentice-Hall, 1962), p. iii. See also M. W. Lifson and M. B. Kline, "Design: The Essence of Engineering," in *Report, Educational Development Program, EDP 5-68* (Los Angeles: Department of Engineering, University of California, Los Angeles, 1968).

19. Layton, *The Revolt of the Engineers,* pp. 26–30, 39–43, 49 n, 51 n, 80, 89.

20. Walter G. Vincenti and Nathan Rosenberg, *The Britannia Bridge: The Generation and Diffusion of Technological Knowledge* (Cambridge, Mass.: MIT Press, 1978); Vincenti, "The Air-Propeller Tests of W. F. Durand and E. P. Lesley," and "Control-Volume Analysis: A Difference in Thinking between Engineering and Physics," *Technology and Culture* 23 (April 1982): 145–74. See also Vincenti, "Technological Knowledge without Science: The Innovation of Flush Riveting in American Airplanes, ca. 1930–ca. 1950," *Technology and Culture* 25 (July 1984): 540–76.

21. Octave Levenspiel, "The Coming-of-Age of Chemical Reaction Engineering," *Chemical Engineering Science* 35 (1980): 1823.

22. Ibid., p. 1825.

23. Ibid., p. 1828.

24. Olaf A. Hougen, one of the principal founders of the modern analytic style of chemical engineering, advocated a more fundamental approach based on "molecular concepts and transport properties," but found physicists not willing to help. His rather disgruntled remarks only mean that he (and other chemical engineers) had to do their own work, and could not expect physicists to do it for them. He warned about carrying science too far, and also emphasized the engineering role in design, construction, and operations. These comments indicate a moderate amount of friction, reflecting different values and priorities established by different communities; they do not indicate a state of war or antagonism. See his "Seven Decades of Chemical Engineering," *Chemical Engineering Progress* 73 (January 1972): 98, 102–3.

25. See for example, Donald S. L. Cardwell, *From Watt to Clausius. The Rise of Thermodynamics in the Early Industrial Age* (Ithaca, N.Y.: Cornell University Press, 1971).

Some Observations on the Historiography of the Industrial Revolution

ROBERT P. MULTHAUF

"The Industrial Revolution" would rank high on any list of overworked topics in the history of technology. But is it not the phrase, rather than the subject, that has been overworked?

It was apparently coined by Arnold Toynbee, and published (post-humously, in 1884) in his famous "Lectures on the Industrial Revolution of the Eighteenth Century in England." The phenomenon in question was essentially English, covering several decades—now seen as roughly the seven from 1760 to 1830 during which certain of the "useful arts," notably the making of textiles and ironfounding, were mechanized, underwent an enormous expansion, and gave Britain a crucial advantage over its economic and political rivals.

Those rivals had been attempting to explain the phenomenon since it had become apparent, about a half-century before the appearance of Toynbee's book. They do not appear to have succeeded, but by 1884 it did not matter, for the "problem" had ceased to exist. The industrial revolution, whatever it was, had swept over the Western world, and England had lost its advantage. Toynbee's book initiated a historical investigation of the phenomenon. So the Industrial Revolution ceased to interest practical men and was left to the historians, who were little attracted to it before the appearance in the early twentieth century of specialists in economic and social history and, still later, in the history of technology. Taking up what seemed virtually the only aspect of the history of technology considered to be of more than antiquarian interest, these new historians pointed out the subjective character of the older literature, which focused on a supposed peculiar genius embodied in English "engineers" or "emminent mechanics," and advanced instead explanations typical of twentieth-century notions of economy or society.

In this essay I have attempted to discover what light can be cast on the subject by a consideration of the terminology; for whatever one may say of the history of technology and of "The Industrial Revolution," their peculiar terminology long predates either.

42

TERMINOLOGY

What were the "useful arts?" Mechanics? Technology? Who was the engineer? As so often, one must begin with Plato and Aristotle, and their essays on classification.[1] *Episteme* (knowledge; in Latin, *scientia*), according to the former, comprises *physis* (nature, the term scarcely signified our "physics" before the nineteenth century), logic, and ethics. Aristotle's more complicated classification began with three forms of knowledge—theoretical, practical, and productive—and then subdivided each of these. Productive knowledge comprised the fine arts and "techne."

What was techne? It was Prometheus's gift to humanity[2] or, more specifically, to medicine (according to Hippocrates, as well as Aristotle), agriculture (Philo Judaeus), building (Vitruvius), warfare (Plato), and the crafts such as textiles and shoemaking (Plato).[3] Education was also techne, according to the Sophists, but this was denied by Plato, who held only those crafts that involved thought, insight, and measurement to be eligible. Shoemaking could qualify, but education (presumably as practiced by the Sophists) could not.[4]

The superior practitioner of techne—the master craftsman—was recognized by name. Plato called him *cheiroteknon,* Aristotle called the "wiser" craftsman *architekton,* and Vitruvius, himself an architekton, said that the "clever" craftsman was possessed of *ingenium*—he was ingenius, perhaps even possessed of genius.[5]

To Plato and Aristotle the whole subject was peripheral, and even distasteful, for they largely shared the ancient attitude typified by Cicero's contrast of the seven *artes liberales* with unspecified artes, which were "minores, sordidae, or mechanicae."[6] But his comprehensive interest in classification obligated the ancient intellectual to concern himself with such matters.

In the phrase just quoted from Cicero we also encounter the term *mechanicae*. Herodotus has used the term in reference to a machine believed to have been used in the construction of the pyramids. It was used by those responsible for the *deus ex machina* so useful in the Greek theater and by the authors of the popular pseudo-Aristotelian text called *Problemata mechanica.*[7]

CLASSIFICATION

Peripheral or not, these cogitations, like so many codified by Plato and Aristotle, survived and resurfaced with the revival of interest in intellectual matters. In Islam, al-Farabi (d. 950), the most eminent Aristotelian of the time, became addicted to classification, and made some personal contributions, adding disciplines, such as alchemy, that had been unknown to the early Greek philosophers.[8] And when ancient writings and the writings of al-Farabi became available in Latin in the twelfth century, classification became a preoccupation of the Latins.

One of these, Hugh of St. Victor, discussed classification in his *Didascalicon,* where he modifies Aristotle (not very helpfully) by denoting the theoretical, practical, and productive as subdivisions of philosophy (philosophia), rather than of science (episteme); and by adding a fourth, the mechanical, with reference to the useful arts or techne. He names these arts, which are (appropriate to tradition) seven: weaving, smithing, navigation, agriculture, hunting, medicine, and the art of the theater.[9]

In putting weaving and smithing first among the parts of mechanical philosophy, Hugh would appear to have previewed the Industrial Revolution. But Hugh was interested in philosophy, not technology. So were his successors, who had by the end of the middle ages developed treelike outlines (*arbor scientiarum*) that grew into great oaks in the course of the sixteenth and seventeenth centuries. At the root lay the supposed divisions of knowledge of Plato and Aristotle, and among the branches we may encounter almost anything, from happiness to pyrotechny and military architecture.[10] The cultivation of these trees, moreover, fell increasingly to academicians who favored the minutiae of school philosophy and restored the mechanical arts to their less conspicuous place as a miscellaneous category.

Perhaps in response to this trend, we encounter in the sixteenth century the classification (if this may be said of hierarchial order) of occupations, rather than disciplines. A book called *Panoply of the Illiberal, Mechanic, or Sedentary Arts,* published in 1568, lists 113, beginning with what one presumes to be sedentary occupations, Pope, Cardinal, Bishop, Priest, and Monk, and ending with tapestrymaker, usurer, glutton, jester, and idiot. Notwithstanding, the majority listed were practioners of the "mechanic arts."[11] So were the "professioni" listed in Tomassino Garzoni's *Piazza Universale* a generation later, some 500 in number, beginning and ending much as had its predecessor, but packed with all manner of peculiar professioni.[12] The following list, picked from his index, may help explain how Garzoni managed to find so many:

> alchimisti, aruspici, astrologi, auguri, geomanti, incanatori, machinatori da oro (gold speculators), malefici, nigromanti, prestigiatori, meritrici and ruffiani (prostitutes and pimps), pirati, banditi, contrabandieri, professori d'enigmi, sibille, superstitiosi, venefici (poisoners), heritici and inquisitori, and otiosi de piazza.

These appear in various groupings, some toward the beginning of his hierarchy, which it seems is not to be taken too seriously. At least one hopes so, for the final profession listed is that of poet.

"Mechanici" are also included in Garzoni's list, where they are discussed in a historical context going back to Tubal Cain.[13] Garzoni deplores the contemporary tendency to use the term mechanic in a derogatory sense, and associates the field with eminent men, particularly with Archimedes, and with practitioners of various arts: builders; the makers of fortifications, war machines, and clockworks; inventors of pneumatic and hydraulic devices; and users of such commonplace "instruments" as the sail, ax, and drill.

Thus Garzoni, despite his tendency to fill the landscape with professional grotesques, seems to bring the practitioners of the useful arts back into the common fold. In fact, the mechanici had already found a common fold—perhaps *denominator* would be a better word—in a new "useful art," that of the firearm. The first cannon appeared in Europe about 1326, and its improvement aroused great enthusiasm among the artist craftsmen of the Italian Renaissance and among the bellicose princes who were their patrons. Among the early printed books (incunabula) appeared the *De re militaria* of Roberto Valturio. The unprinted *De architetture civile et militare* of Francesco di Giorgio Martini was written about 1475 and was soon followed by Leonardo da Vinci's various writings on military inventions.

One authority on the history of firearms has claimed that all possibilities in their size and variety were already exhausted by these early authors[14]—a claim that a glance at their books tends to support. But none of them were in fact exclusively concerned with military machines. This was also the time of the recovery and popularization of such ancient technical writings as those of Heron of Alexandria and Vitruvius, and with them the idea of mechanical invention. Out of their amalgamation with what I will call the "gun books" came the celebrated renaissance "machine books," culminating in the multivolume *Theatrum instrumentorum et machinarum* published by Jacob Leupold in the 1720s and the 1730s.[15]

ENGINEERING

If the titles of the machine books reveal the importance of the term *machine,* it does not follow that their authors were called "mechanics. They were generally called "engineers," as is most strikingly shown in the case of Leonardo, who was so called by all of his several patrons.[16] The term appears to derive ultimately from the ingenium possessed by Vitruvius's superior craftsman and to imply more than the conduct of routine business. Garzoni's professioni also include the engineer, whom he differentiates from the mechanic in that the latter works both with the hand and the mind, while the engineer works only with the mind.[17] Thus Garzoni appears to have had in mind not the inventors but the users of firearms, the directors of artillery who were becoming increasingly important in the military establishment. These engineers were not, like the mechanics, a collection of ingenious individuals. They were a new class, experts in the use of firearms, tacticians and strategists. And they were busy, for after its early successes the cannon was increasingly frustrated by improvements in the art of fortification—as the historian Charles Oman put it, "the spade got the mastery of the cannon ball."[18] But the military engineer, it seems, was in an unassailable position, for he had charge of both departments, assault and defense. Indeed he had employment that was not only independent of the relative fortunes of the cannonball and the spade, but even of the military art altogether. For his art was susceptible to extension toward what we call "civil engineering." When, toward the end of the sixteenth century, the Dutch wars with Spain were tapering off, the Dutch engineers found an appropriate occupation in dike building. In 1600 Simon Stevin, the most distinguished of them, established at Leyden a school specifically aimed at teaching what he called "engineering." It was the first engineering school.[19]

TECHNOLOGY IN ENGLAND

England had indeed been backward in the "mechanic arts," and had depended extensively on foreigners in such traditional industries as mining, ironfounding, saltmaking, gunpowder, glass, and textiles, some of which they virtually founded.[20] Some British steel works were still manned by foreigners in the late seventeenth century, and it has been said that it was still "common" at that time for English mineral enterprises to be under the control of the Dutch.[21] To a degree this practice was even true in agriculture as we see in the mid-seventeenth century when Robert Child called for the settlement in Ireland of "Dutch experts in modern agricultural methods."[22] Dickinson has claimed that England "drew ahead gradually" from the sixteenth century in the textile industry, but his principal evidence is the career of William Lee (1563–1610), inventor of the stocking frame, who took his invention to France because it was not appreciated at home[23]—testimony to British ingenuity but hardly to the superiority of the British textile industry.

Apparently, none of the continental machine books was translated into English. The writings of Ramelli, Zonca, Besson, and others made an appearance of sorts, when they were partly incorporated into John Wilkins's *Mathematical Magic* of 1648. But Wilkins's book was trivial by comparison. The contents of the machine books came in more effectively with immigrants. Such were the Caus family, who appeared about 1590, to continue—through its most illustrious members, Isaac and Salomen—careers as inventors and builders of novelties, especially fountains, for the landed gentry. That the natives were not on the whole much interested is suggested by the fact that Isaac's *Nouvelle Invention de Lever l'Eau*, although published in London, was written in French. About 1604 the legendary Cornelius Drebbel appeared in England, where he was to pursue an eccentric career as an inventor reminiscent of the Italian virtuosi of the fifteenth century. He even enjoyed the favor of the English court, which in general had lacked the interest in the company of machine inventors that was typical on the continent.[24]

The Caus family was Huguenot, from Dieppe. Drebbel was Dutch. Representatives of the Low Countries, as has already been implied, were remarkably prominent in English technology in the seventeenth century. The first pumping engine for supplying water at London Bridge was installed in 1682 by a Dutchman, Peter Morris.[25] Robert Child's call for Dutch experts in agriculture may have called forth Gabriel Plattes, probably Dutch, whose *Description of Infinite Treasures* (1639) has been characterized as a pioneering English work on agriculture.[26] Plattes also published a small book on mining in 1679, with the assistance of Arnold Boate, another Hollander, whose brother, Gerard, had already published a book descriptive of, among other things, the economic geology of Ireland.

The most prestigious visitors, however, were dike builders, who were specifically called "engineers." In 1582, after centuries of amateurish effort to construct a harbor at Dover, the English turned to plans submitted by men from the Low Countries, such as Humphrey Bradley, who had previously been employed in France as "Maitre des Digues" (dikes).[27] Bradley subsequently became involved in the draining of the Fens, the great British public work of the seventeenth century. It was directed by the "Commissioners of Sewers," who appear to have been a bureaucracy assembled from the nobility. The work was contracted to various "adventurers," mostly of noble origin, but actually conducted in large part by Dutchmen, of whom the most famous was the "engineer," Cornelius Vermuyden.[28]

Britain was not without native "engineers," although their designation as such was mostly posthumous, perpetrated by the celebrated Samuel Smiles, whose *Lives of the Engineers* appeared in 1861–62 and was followed in 1879 by his *Industrial Biography*. Smiles's books were a kind of history of "engineering" in Britain, built around biographies. They originated in historical reflections on bridge and canal builders, which he subsequently broadened to include ironworkers and "toolmakers." Smiles was idiosyncratically selective, but he can be supplemented by Ludwig Darmstaedter's *Handbuch zur Geschichte der Naturwissenschaften und der Technik* (1908), which is more concerned with invention and scarcely at all with civil engineering. Taken together, these two authors give us a rough picture of the native practioners of technology of Britain.

Between them, Smiles and Darmstaedter provide the names of about fifty-five important British technologists active between 1500 and 1750, by which date British supremacy in the most important technologies was assured.[29] Precision is impossible, as neither of these publications fulfills the requirements of a biographical dictionary,[30] and some of the technologists are barely known beyond their connection with some particular event or invention. But some generalizations can be made. They were neither urban, nor bourgeois. Most of them were of rural origin, and about one-third were of the nobility. It is likely that all could read English, but we have specific information on only about one-half, and these are equally divided between those who attended college and the self-trained. Indeed, one-quarter belonged to the Royal Society (excluding those who lived too early)—but nearly all of these were makers of scientific instruments, a subject of passionate interest to the gadget-loving academicians.

In any case, all of them were probably literate.[31] What would they have read? Assuming that most of them were restricted to the English language, they would have been limited to the following books, most of which have already been mentioned:

> 1635   John Bate, *Mysteries of Nature and Art.*
> 1639   Gabriel Plattes, *Description of Infinite Treasures.*
> 1648   John Wilkins, *Mathematical Magick.*
> 1652   Gerard Boate, *Ireland's Natural History.*
> 1659   R. d'Acres, *The Art of Water Drawing.*
> 1663   Marquis of Worcester, *Century of Inventions.*
> 1683   Hugh Plat, *The Jewel House of Nature.*
> 1683   Joseph Moxon, *Mechanic Exercises.*

In Charles Webster's recent massive study of science in Britain during the two decades before 1660, an elaborate argument has been made for the "baconianism" of most of these authors[32]—for the influence, that is, of Francis Bacon's advocacy of the application of "science" to the arts. Yet, excepting Moxon, none of them so much as mention Bacon, and Moxon's reference is cursory, a nod to what had become the fashion by the late seventeenth century.[33] Indeed, these books add further evidence for the influence of the Netherlanders, for as many as half of the authors may have been Dutch immigrants, and even that indubitable Englishman, the Marquis of Worcester, stressed the importance to his work of his assistant for thirty-five years, Caspar Kaltoff, a Dutchman.[34]

In any case, I think it doubtful that these books, with the possible exception of Moxon's, could have had significant influence on the technologists mentioned by Smiles and Darmstaedter. With that exception, these books contain virtually nothing that would have been of use to the civil engineer, ironworker, or toolmaker. The books of d'Acres and Worcester might conceivably have influenced the inventors of the steam engine, but could hardly have inspired such inventions as the stocking frame, flying shuttle, washing machine, writing machine, or water closet. For all of these, the example of Cornelius Drebbel seems more probable.

Ultimately the supposition that Bacon influenced the rise of Britain to supremacy in technology appears to rest on the assumption that his advocacy of the *application of science* to techology, and the response of the Royal Society to that advocacy, must have had some result.[35] Moreover ten or eleven of our forty-seven technologists (six lived too early) were Fellows of the Royal Society.[36] But the most influential were probably Samuel Morland, who rather resembles a spiritual descendant of Cornelius Drebbel, and John

Smeaton, probably the most intellectual British technologist of his time. If Bacon inspired them, it was at very long range, for both were active in the last half of the eighteenth century, and indeed neither seems particularly "baconian" either ideologically or empirically. The appearance among the membership of the Royal Society of persons active in technology seems merely to reflect the miscellaneous character of that body and its particular fascination with scientific instruments.

Nonetheless, by the early eighteenth century the native technologists of Britain began to exhibit a confident virtuosity. It was the same quality that had earlier characterized the authors of the Continental machine books, but with significant differences. The English were less concerned with the ingenious showpieces that appealed to royalty. Their concerns are already evident in Moxon's *Mechanic Exercises,* which, notwithstanding its utilization of the Continental books, is basically concerned with sober topics such as carpentry and ironworking. And the British "engineers" had little connection with the Crown, or even with the government—not to mention the idealized laboratories of Bacon's New Atlantis. Even in civil engineering projects, they usually appear as private contractors, often working on private projects.[37] By comparison with those in the Continental machine books, their inventions, a number of which have been mentioned, seem to be directed at the mundane needs of society rather than the flamboyant projects characteristic of governments.

Bacon might have cheered them on, but did his writings inspire them? England's emergence as a technological and industrial colossus at the end of the eighteenth century is not easily connected with the preoccupations of the classifiers of the sciences, with Bacon's prescriptions for their reformation, with the injunctions of the Royal Society, or even with the trades as they were described in books. Ultimately it probably rested on the challenge of the wood shortage, the substitution of coal, with which Britain was uniquely supplied, and the discovery of the coking process by Dud Dudley. Are these accomplishments not more likely to have been inspired by the problem-solving attitudes of the foreign engineers who swarmed to England during the previous century? Is it not likely that Dutch custom made the British mechanic an engineer—and the steam engine an engine instead of a machine?

That a technologically underdeveloped society, which Britain clearly was through most of the seventeenth century, would blossom into leadership after fertilization by advanced foreign techology was probably not unprecedented. And it was not to be unique. Much the same happened a century later in the United States and two centuries later in Japan.

NOTES

1. The relevant writings of Plato and Aristotle are quoted in Edward Grant, ed., *A Sourcebook of Medieval Science* (Cambridge, Mass.: Harvard University Press,

1974), p. 53. See also Richard McKeon, "Philosophy and Theology, History and Science, in the thought of Bonaventure and Thomas Aquinas," *Speculum* 36 (1975): 387–412.

2. "Pantechna" (all arts); Aeschylus, *Prometheus Bound.*

3. Aristotle, *Politics* 3. 11, and Hippocrates, *On Ancient Medicine* 1. 1 (medicine); Philo Judaeus, *On Husbandry* (Philo Judaeus, "On Husbandry," in *Philo,* ed. F. H. Colson [Cambridge, Mass.: Harvard University Press, 1958], 3:108–9) (agriculture); Plato, *Laws* 919–21 (warfare); *Repulic* 495 A–E (the crafts).

4. See Werner Jaeger, *Paideia* (New York: Oxford University Press, 1945), 1:3; 2:129–30, 306; 3:66, 188.

5. Plato, *Republic* 590C, *Philebus* 55–56. Aristotle, *Metaphysics* 980–81. A similar distinction was made for agricultural workers in Philo Judaeus, "On Husbandry," 3:109. Vitruvius, *De architectura* 1. 1. 1.

6. Cicero, *De officio* 1. 42. 150.

7. Herodotus, *History* 2. 125. The deus ex machina is mentioned, for example, in Plato, *Cratylus* 425D. In addition to the pseudo-Aristotelian *Problemata mechanica,* there is a *Mechanicae syntaxis* by Philo of Alexandria and a *Mechanica* by Hero of Alexandria. See also H. M. Klinkenberg, "artes liberales/artes mechanicae," in *Historisches Woerterbuch der Philosophie,* vol. 1, ed. Joachim Ritter (Basel: Schwabe, 1971).

8. Al Farabi's book called by the Latins *De scientiis* was translated in the twelfth century. It has been published in German translation in Clemens Baumker, ed., "Alfarabi: Über den Ursprung der Wissenschaften (De ortu scientiarum)," *Beitraege zur Geschichte der Philosophie des Mittelalters* 1, 19, no. 3 (1916).

9. Hugh of St. Victor, *Didascalicon,* 2. 21, trans. Jerome Taylor (New York: Columbia University Press, 1961), pp. 74–75.

10. Cf. N. H. Steneck, "A Late Medieval Arbor Scientiarum," *Speculum* 50 (1975): 245–69, and Giorgio Tonelli, "The Problem of the Classification of the Sciences in Kant's Time," *Rivista Critica di Storia Filosofia* 30 (1975): 243–94.

11. Hartman, Schopperum, ΠΑΝΟΠΛΙΑ *omnium illiberalium mechanicarum aut sedentariarum artium genera continens* (Frankfurt, 1568), illustrated by Jost Amman, as was *Eygentlilche Beschreibung aller Staende auf Erden* (Frankfurt, 1568), where poems by Hans Sachs accompany the illustrations. These appear to be two versions of the same book.

12. Tomassino Garzoni, *Piazza universale* (Venice, 1665), discorso 13, 39–42, 63, 75, 117–18, 138, and 144. There are a total of 154 discorso.

13. Ibid., p. 539.

14. Napoleon III, "Du passe et de l'avenir de l'artillerie," in his *Oeuvres* (Paris, 1856), 4:59.

15. Juanelo Turriano, *Libros de los ingenios y maguinas* (in manuscript, see Ladislao Reti, "The Codex of Juanelo Turriano," *Technology and Culture* 8 [January 1967]: 53–66); Jacques Besson, *Theatrum instrumentorum et machinarum* (Lyons, 1578); Agostino Ramelli, *Le diverse et artificiose machine* (Paris, 1588); Vittorio Zonca, *Novo teatro di machine et adificii* (Padua, 1607); Heinrich Zeising, *Theatrum machinarum* (Leipzig, 1613–55); Faustio Veranzio, *Machinae novae* (Venice, 1595); Salomen de Caus, *Les raisons des forces mouvantes avec diverses machines* (Frankfurt, 1615); Giovanni Branca, *La machine* (Rome, 1629); Caspar Schott, *Mechanica hydraulica-pneumatica* (Frankfurt, 1657); G. A. Boeckler, *Theatrum machincarum novum* (Nurnburg, 1661); Jacob Leupold, *Theatrum machinarum* (Leipzig, 1724ff.)

16. Ludovico Sforza called Leonardo his "ingenarius et pinctor"; Cesare Borgia called him "architecto et engegnero generale"; King Louis XII of France called Leonardo "paintre et ingenieur ordinaire." Documents relating to Leonardo's burial,

in France in 1519, call him "peintre et ingenieur et architecto du Roi, mescanicien d'Estat." See the biography in vol. 8 of the *Dictionary of Scientific Biography.*

17. Garzoni, discorso 107, pp. 556–61.

18. Charles Oman, *A History of the Art of War in the 16th Century* (New York: Dutton, n.d. [1937]), p. 223.

19. E. J. Dijksterhuis, *Simon Stevin* ('s Gravenhage: M. Nijhoff, 1943), p. 14.

20. Cf. Rhys Jenkins, "The Rise and Fall of the Sussex Iron Industry," *Transactions of the Newcomen Society* 1 (1920): 16–33, and idem, "Observations on the Rise and Progress of Manufacturing Industry in England," ibid. 7 (1926): 1–15; B. C. Halahan, "Chiddingford Glass and Its Makers in the Middle Ages," ibid. 8 (1926): 188–95; H. W. Dickinson and A. A. Gomme, "Netherlands' Contributions to Great Britain's Engineering and Technology to the Year 1700," *Archives Internationales d'Histoire des Sciences* No. 3 (April 1950):356–77.

21. Rhys Jenkins, "Notes on the Early History of Steel Making in England," *Transactions of the Newcomen Society* 3 (1922): 16–23; Charles Webster, *The Great Instauration* (New York: Holmes and Meier, 1975), p. 334.

22. Webster, *The Great Instauration,* p. 432.

23. H. W. Dickinson and A. A. Gomme, "Some British Contributions to Continental Technology, 1600–1850," *Archives Internationales d'Histoire des Sciences* No. 16 (July 1951): 706–22. They also mention John Kay (1714–79?), inventor of the flying shuttle, who also took his invention to France.

24. Drebbel, who marched in the funeral procession of James I (1625), was described by C. F. de Peiresc as "Ingenieur du Roi d'Angleterre" (L. E. Harris, *The Two Netherlanders* [Cambridge: Heffers, 1961], pp. 126, 194). J. U. Nef has remarked on the general lack of interest of British monarchs in mechanical virtuosi (*Industry and Government in France and England, 1540–1640* [Ithaca, N.Y.: Cornell University Press, 1957], p. 103).

25. Rhys Jenkins, "Notes on the London Bridge Waterworks," in his *Collected Papers* (London, 1936), pp. 131–40.

26. *Dictionary of National Biography.* See also Webster, *The Great Instauration,* pp. 471–72.

27. Harris, *The Two Netherlands,* pp. 11–18.

28. Ibid., pp. 106, 166–67. H. C. Darby, *The Draining of the Fens* (Cambridge: Cambridge University Press, 1940), pp. 28–42, 59–60.

29. From Smiles I have the following names (in alphabetical order): Joseph Bramah, James Brindley, Heny Cort, Abraham Darby, Dud Dudley, William Edwards, Benjamin Huntsman, John Metcalf, Hugh Myddelton, (John) Perry, John Rennie, John Roebuck, John Smeaton, and Andrew Yarranton. From Darmstaedter I have: John Astbury, (Robert) Barker, Edward Barlow, Henry Beighton, Thomas Bolsover, P. Bowde, R. Broke, Wm. Caxton, —. Clement, J. Cumberland, Abraham Darby, Dud Dudley, James Ferguson, Wm. Ged, Thomas Godfrey, George Graham, R. Hage, J. Hardley, J. Harrington, John Harrison, H. Haskins, Robert Hooke, R. Houghton, J. Hull, Benjamin Huntsman, R. Jennings, John Kay, John Law, Wm. Lee, Hugh Mackay, Henry Mill, Samuel Morland, Thomas Newcomen, — Pashley, J. Payne, Phineas Pett, Hugh Platt, Humphrey Potter, John Roebuck, Thomas Savery, James Short, (Edward) Southwell, J. Tyzach, E. Ward, and John Wyatt.

30. Smiles's works are not biographical dictionaries; they are a kind of history of technology in Britain, structured around biographies. Many persons are mentioned, but only a few—I estimate fourteen born before 1750—are given major treatment. Darmstaedter's *Handbuch* deals not with biographies, but with "events."

31. Widespread ability to read English is stressed in J. W. Adamson, "The Extent of Literacy in England in the 15th and 16th Centuries: Notes and Conjectures,"

Library (ser. 4), 10 (1929–30): 163–93. It is also emphasized for the seventeenth century by Webster, *The Great Instauration,* pp. 207–17.

32. Webster, *The Great Instauration,* passim.

33. It appears that only in the second edition did Moxon discover that his work was in accord with the prescriptions of "Lord Bacon." (Joseph Moxon, *Mechanick exercises,* 3d ed. [1703]; reprint, New York: Praeger, 1970, p. xiv and preface.)

34. Henry Dircks, *The Life, Times and Scientific Labors of the Second Marquis of Worcester. To which is added a reprint of his Century of Inventions, 1663* (London: Quaritch, 1865), p. 383 (the Marquis' preface).

35. It appears to me that this criticism can be made of Webster's *The Great Instauration,* to which I have made frequent reference, and also of R. K. Merton's classic, *Science, Technology and Society in 17th-century England* (first published in *Osiris* 4 [1938]: 360–632). The work was reprinted, with a new introduction, by Howard Fertig (1970) and by Humanities Press, 1978. The assumption appears to rest on a failure to attempt seriously to differentiate science from technology. These books tend to assume that what has been demonstrated for science also applies to technology.

36. As indicated in *The Record of the Royal Society,* 4th ed. (London and Edinburgh: Morrison & Gibb, 1940).

37. See Smiles's biographies of Hugh Myddelton, John Metcalf, William Edwards, and James Brindley.

Technical Journals and the History of Technology

EUGENE S. FERGUSON

Periodical literature supplies a significant portion of source material in the history of technology. It frequently contains the only contemporaneous account we can find of a particular event or controversial issue. Yet we often know surprisingly little about the circumstances surrounding those articles, news items, or editorials that we depend upon in our research.

Because articles are often unsigned and because mastheads usually yield little direct information regarding an article's context, it is easy to assume that all editors have been concerned primarily with conveying to their readers a balanced, objective view of the topics treated. Professionally trained historians are usually more familiar with academic journals than they are with technical journals. They know the ideal of ethical standards in academic journals: fairness, willingness to include a variety of viewpoints and the intent to advance and encourage intellectual debate. Melvin Kranzberg exemplified this ideal in his creation and editing of *Technology and Culture*.

On the other hand, nineteenth-century editors of technical journals, who set standards that still prevail, envisioned a very different role for themselves. That difference can be readily demonstrated by taking a closer look at nineteenth-century journals. Those editors came to their work with strong viewpoints and urgent agendas. Some were uncritical advocates for their industry, taking every opportunity to lobby on its behalf. Others were consistently critical as they reported events and ideas; their journals apparently were successful because they brought intelligence and freshness to the pages they published. Still others cultivated the new field of popular science and exploited the popular myth of invention as an easy road to riches. Another group had particular avocations or hobbies that transformed them into crusaders or, in the biblical sense, prophets. Technology may not be autonomous, but its boosters and barkers are.

In general, technical editors used their journals to formulate opinion and mold thought. With their ability to have their words read by people who possessed the authority to make or resist changes, they seldom left the world as they found it. To use Leonard Reich's perceptive phrase, the editors "wielded little authority but possessed great power."[1]

In the series of vignettes of editors and publishers that follows, I have tried to illuminate some of the strengths and weaknesses that may be observed in technical journals of the nineteenth century and to supply a sense of the significant information and insights that lie between the lines of technical journals of any age. The essay closes with observations on procedures and problems of using technical periodicals as historical sources.

Some years ago, I wrote an article about John Ericsson's "caloric" ship of 1852.[2] Of the eighty-seven references in that article, sixty-six were to periodicals. The rest were to patents (seven), to nineteenth-century handbooks and textbooks (six), steamboat histories (three), an Ericsson biography (one), and Ericsson papers in a historical society (four). I was interested chiefly in the ship's enormous hot-air engines, designed by Ericsson, who believed wrongly that his engines were as powerful as steam engines and inherently more economical. Ericsson invited newspaper reporters to a carefully staged trial trip, replete with a banquet, speeches, resolutions, and toasts. The trial was technically a failure, with the engine delivering about one-eighth as much power as would a comparable steam engine, but the failure was neither evident nor relevant to most of the company: newpapers embraced Ericsson's conviction that "caloric"—hot air—would supersede steam. However, as the reporters came ashore after the trial, a skeptical voiced shouted "Vive la humbug." Because the report of the trial in *Scientific American* doubted the extravagant claims of Ericsson and because its writer claimed to have been an uninvited guest at the trial, I assumed that the dissident writer was Orson Munn, the 28-year-old publisher of *Scientific American,* and that he was, moreover, the discourteous outspoken skeptic.

In due course, I received a letter from Michael Borut, who was working on his doctoral dissertation at New York University, asking me how I knew Orson Munn was present at the trial of Ericsson's ship. I admitted that I did not know positively but had deduced his presence from what I thought I knew about *Scientific American.* I had assumed that Munn was his own chief reporter as well as editor, an assumption that the journal itself has never yet hinted was in error.

Eventually I learned from Borut's dissertation, "The *Scientific American* in Nineteenth Century America," that Munn was not his own chief reporter, nor was he even editor of his journal. I learned a great deal more from Borut, whose study of one of the most influential technical journals of the century was solidly based on a 64-year run of Munn's personal diaries.[3] However, the most lasting lesson I learned is that deductions about a journal based on knowledge derived only from that journal are likely to be wrong.

SCIENTIFIC AMERICAN

Rufus Porter, who founded *Scientific American* in 1845, had little to do with its success or influence because he sold it less than a year after it commenced publication to Orson Munn and Alfred Beach, who were, respectively, twenty-one and nineteen years old at the time. Munn and Beach knew each other at Monson Academy, near Springfield, Massachusetts; Beach was the son of Moses Beach, owner of the New York daily *Sun,* in whose building "Munn & Co." had its first office.[4] Porter did, however, supply the magical name, which gave a powerful boost to the notion that science, a term of higher status, could and should be used interchangeably with technology and invention. Porter's subtitle gave a more accurate notion of the aim and scope of the journal, which did not change after the new owners took charge: *The Advocate of Industry and Enterprise, and Journal of Mechanical and Other Improvements.*

The extraordinary success of *Scientific American* in the nineteenth century was due to Orson Munn's ability as an entrepreneur rather than editor. (Alfred Beach dropped out for eight years, buying back into the firm in 1854.) It was Munn and the editor he hired, Robert Macfarlane, a Scottish immigrant, who worked out the course that the journal followed.[5] In 1847, Munn & Co. as publishers of *Scientific American* opened its own patent agency, designed to act as intermediary between individual inventors and the U.S. Patent Office in Washington, D.C.[6] The editorial content of the journal was subordinated to the aim of building a business as patent solicitor but, judging from the journal's growing circulation, readers were apparently quite happy to read whatever Macfarlane wrote or borrowed from other journals or books.[7]

A typical weekly issue (12 January 1850) included a few one-column "how-to" pieces (how to color sheepskins as doormats; how to tan leather; a reader's letter giving practical information about waterwheels); a serial installment of the editor's "History of Propellers and Steam Navigation," which was eventually published as a book; numerous short pieces, most of them borrowed from newspapers and books, on topics such as the destruction by seventy-five Detroit residents of a railroad they considered a nuisance, on dress in Japan, on the value of parsnips as animal food, on English piracy of American inventions, on British maps and government surveys, on blowing up marine wrecks with electricity, and many others. An editorial on "Reading on Winter Evenings" was in the self-help tradition of the early Victorian era and included the exhortation, "Young man, whatever others have been, you can be, but not without effort—conscious, unwavering effort."

Nearly half of a typical issue was devoted directly to patents and how to get them. Recent patents, procured for the inventors by Munn & Co., were illustrated and explained in detail: a new smut-removing grain mill, a new way of making lead shot, a pencil case, a new rotary engine. A running list of patent claims, issued weekly by the U.S. Patent Office, was published promptly; a report concerning the Patent Office by the Secretary of the Interior was given ample space and critical examination and commentary.

Several dozen short replies to readers' letters helped build the image of *Scientific American* as a knowledgeable, independent, and trustworthy source of information and opinion. "A.G., of Ohio" was told, "We do not see how you could be refused a patent, for your idea is *new,* and it is certainly *useful*—the two essentials requisite to secure a patent." "J.J., of N.Y.," on the other hand, learned that "It does not appear that you have discovered any new principle in atmospheric churns. The same dash has been shown us within a few weeks. Independent of this, the modifications are not patentable as we view them."

Finally, a prominent advertisement addressed to "Inventors and Mechanics" extolled the virtues of "The Best Mechanical Paper in the World!" and urged readers to subscribe.

Munn & Co. supplied properly written specifications and suitable mechanical drawings, made searches in the Patent Office for earlier patents that might interfere with the patent being sought, reworked and resubmitted rejected applications, and in general offered inventors the benefit of familiarity with the official requirements and informal practices of the Patent Office. Except for the first opinion on patentability, none of Munn's services were free. Charges, while not extravagant, were sufficient to cause Orson Munn to write in his diary in 1852 "how singularly and extravagantly we are prospered."[8]

Prospective clients of the Munn & Co. Patent Agency were encouraged by the success of other readers in obtaining patents and enticed by attractive supplementary booklets that told readers "How to Get Rich"—by obtaining patents, of course. Everybody can invent, said one booklet, "without distinction of race or color . . . women or minors . . . foreigners as well as citizens. All that is necessary," it continued, "is *to think;* not profoundly, but in a simple, easy way, which every one can do."[9] Most, but not all, of the patents obtained through the agency certainly exhibited an absence of profound thought and many could be described as lacking in common sense, but patents were secured in large numbers.

Scientific American soon became the leader and chief member of the patent lobby, bringing whatever pressure it could command for more examiners and less rigorous examination of applications. The law that reformed the Patent Office in 1836 gave examiners no clear guidance in their decisions regarding the merits or novelty of an invention. Some of the early examiners were excessively zealous in uncovering "prior art"; in the years around 1850, fewer than half of the applicants were granted patents.[10] When Charles Mason, Patent Commissioner from 1853 to 1857, agreed with the aims of the patent lobby and was diligent in making it easier to obtain patents, *Scientific American* praised him highly. When he ended his term in 1857, Mason reciprocated by writing a glowing testimonial to "Messrs. Munn & Co.," praising their "*promptness, skill, and fidelity* to the interests of your employers"—that is, the patentees. Mason also noted that while he was Commissioner "more than one fourth of all the business of the office came through your hands." Two subsequent commissioners wrote similar letters, emphasizing energy, ability, and fidelity, and remarking upon how "very large" was the business brought to the Patent Office by Munn & Co.[11] In 1860, Judge Mason went to work for Munn & Co. He stayed less than a year, but his name, prestige, connections, and insider's knowledge of the Patent Office enhanced the abilities and status of the largest patent soliciting firm in the world.[12] The employment of a former commissioner by the firm bringing the Patent Office one-fourth of its business also served to blur the boundary between the public Patent Office and a private firm of patent solicitors. The volume of business grew, and in 1868 Munn & Co. claimed that "more than one-third of all patents granted are obtained by this firm." Later claims are ambiguous, but at least a hundred thousand patents had been obtained by the time Orson Munn died in 1907.[13]

Scientific American had a number of imitators, but none made serious inroads into either its size or influence. Henry T. Brown, who had worked in the *Scientific American* Patent Agency since 1849, resigned in 1864 to found the *American Artisan*, a patent journal that used *Scientific American* as its pattern. John W. Combs, who had been an editor at *Scientific American* for twelve years, resigned to become editor of the new journal. *American Artisan* survived for eleven years, and Brown continued his own patent agency for at least thirty years after his journal folded.[14] *American Inventor*, another imitator of *Scientific American*, was published from 1878 to 1907 by the American Patent Agency of Cincinnati, which had branch offices in St. Louis and Atlanta.[15]

According to Munn & Co., the competitors who would gladly take inventors' money were almost legion: "there is scarcely a town of 4,000 inhabitants" that did not have a patent agency or patent attorney. Yet Munn & Co. led the pack by becoming identified with what it called "the universal brotherhood of Inventors and Patentees at home and abroad." Munn's journal boasted that "with the increased activity of these men of genius we have kept pace up to this time, when we find ourselves transacting a larger business in this profession than any other in the world."[16] A few prominent firms of patent attorneys had enviable reputations, but none accounted for more than two or three per cent of all patents issued.[17]

Its large circulation and its ability to encourage its readers "*to think . . .* in a simple, easy way" and become inventors, confirmed the notion that *Scientific American* was, as its editors believed, important to the "scientific students who have found in the *Scientific American* their weekly pabulum."[18] The journal appealed to true believers, children of all ages who were avid in their pursuit of novelty and who equated novelty with progress.

Over time, the journal broadened its editorial scope and published a number of informative and well-illustrated technical accounts of such important projects as the Brooklyn Bridge (opened 1883), the Statue of Liberty

(1886), and the great Ferris Wheel at the Chicago exposition of 1893. Also described in words and pictures were the effects on New York City of the blizzard of 1888. A series on "American Industries," commencing in 1879 and continuing through 1882, typically featured a full front-page wood-engraving showing several views of the particular company being described, a page (two thousand words) or more of text, and two or three small wood-engravings to support the text further. A *Scientific American* artist visited a company's plant and made sketches from which the wood-engravings were made. An official of the company either supplied the text or employed a *Scientific American* writer to do so.[19]

Alfred Beach, who resumed his partnership in the journal during the 1850s, became increasingly active after the departure of Brown, Combs, and Macfarlane in 1864, while Munn relaxed the tight control he exercised in earlier times. Beach suggested the publication of *Scientific American Supplement,* a weekly journal similar to but separate from *Scientific American,* and without the burden of patent promotion. Munn thought it a bad idea but acquiesced; the *Supplement* was launched in 1876 and outlived both Beach and Munn, being published well into the twentieth century. Beach also published, in Spanish, a Latin American edition of *Scientific American.*[20]

In the years of rapid industrialization following the Civil War, technical opportunities and social institutions combined to spread the ideology of progress and the patterns of behavior that made progress possible. Mercantile Libraries in the big cities and the Young Mens Christian Associations in many towns and villages brought polite and useful reading matter, including *Scientific American,* to interested readers and brought together young men who became immersed in ideas and information relating directly to their visions of success in an intensely interesting technical world.[21]

One wonders how idealistic readers viewed the journal's policy of charging for space in the editorial as well as advertising pages. The "American Industries" series was a logical outcome of earlier advertisements of Munn & Co. offering to "send artists to make sketches of manufacturing establishments, with a view to publication in the Scientific American." Not only did manufacturers pay for these image-building articles, but so did any inventor whose patent was illustrated and described in the journal.[22] *Scientific American* was not the only periodical to accept money for editorial promotional pieces, but three important technical journals, at least, were publicly opposed to such a policy. *Engineering News* (later *Engineering News-Record*) in its first volume declared its editorial independence by asserting that "not a line" had been or would be written "in the interest of any manufacturer." *Railroad Gazette* stated weekly in its "Editorial Announcements" that the editors "wish it distinctly understood that we will entertain no proposition to publish anything in this journal for pay, *except in the advertising columns.*" *American Machinist* was as definite. "Positively," wrote its editor, "we will neither publish anything in our reading columns for pay or in consideration of advertising patronage."[23]

Yet no journal had a wider following than *Scientific American* or, probably, a wider influence upon popular conceptions of progress and the popular notions regarding the rewards of patenting. The aura of *Scientific American,* always privy to the latest technical wonder, never at a loss for a confident answer to any query, encouraged readers to suspend critical judgment and to accept without examination whatever issued from the facile pens of its anonymous writers.

Scientific American prepared young America for *Popular Mechanics,* a monthly compendium of engineers' dreams, supplemented by specific instructions on how to build motors and radios and how to modify old Model T cars. *Popular Mechanics* appeared in 1902, published by Henry H. Windsor, 41, an Iowan, graduate of Grinnell College, and already editor of two trade magazines. Windsor had a keen sense of what would hold his readers. In the words of the official fifty-year history of the journal, its editors "looked at the present and future of science, mechanics, and invention, with scarcely a backward glance."[24]

My generation learned not only the latest brainstorms of the illustrators who drew speeding trains, rudimentary snowmobiles, cutaway submarines, and airplanes dropping bombs, but we also had reinforced in our developing minds the myths of a benign technology. A typical piece was one in 1932 entitled "Machines—Masters or Slaves." Yes, wrote the author, in the year when one-fourth of all Americans who wanted to work were without work, machines do replace men, but in the long run machines make new jobs, much less onerous than the ones that have been destroyed. We learned from an article on testing consumer products that things are not as good, "in a technical sense," as they used to be, but that consideration of the vast increase in quantity and variety leads to the "undeniable fact that living itself is better."[25] According to journals like *Scientific American* and *Popular Mechanics,* the ideas and ideology presented by the journals are not dangerous: it is the people who act on them who do the damage.

Whether we applaud or deplore the results of technical progress in the twentieth century, we can place a not inconsiderable share of the praise or blame on the editors of *Popular Mechanics* and similar journals of technical wonders, whose mission has been to keep alive the legacy of *Scientific American,* that pioneer of progress barking.

ENGINEERING

Two perennially successfully English technical journals, *The Engineer* and *Engineering,* were founded in London just ten years apart, in 1856 and 1866, respectively. The first was started by Edward Henley, a young man in his thirties;[26] the second, with whose editor these notes are concerned, was started by Zerah Colburn, an American, also in his thirties, who had an early and abiding interest in steam locomotives.[27] At the age of eighteen, Colburn wrote and published a modest volume entitled *The Locomotive Engine, Theoretically and Practically Considered.* He was employed for two or three years in locomotive works in Boston and Richmond, after which he joined the staff of the *American Railroad Journal* in New York. In 1854, when he was twenty-two, he started his own weekly journal, *Railroad Advocate.* Two years later he sold it to Alexander Holley, also twenty-two, who in the 1860s was to become the leading American engineer of Bessemer steelmaking. In 1857, Colburn and Holley traveled together to Europe to gather material for a large, authoritative book on *The Permanent Way and Coal Burning Boilers of European Railroads,* published in New York in 1858. Soon thereafter, Colburn returned to England and, before the year was out, was editor of *The Engineer.* He stayed for a while, but in 1860 he resigned and returned to New York on the first voyage to America of Brunel's great steamship *Great Eastern.* Almost immediately, he started in Philadelphia a new weekly journal named *The Engineer,* which had no connection with the London weekly he had just left. Five months later, his journal ceased publication for lack of support and Colburn returned yet again to London, where he resumed his editorship of *The Engineer,* a post he held until late in 1864.

During his time in England on *The Engineer,* Colburn published books on steam boiler explosions, the nature of heat, the gas works of London, and locomotive engineering. In London, he joined the Institution of Civil Engineers and the Institution of Mechanical Engineers, and he served as president of the London Society of Engineers. He read papers to those organizations, to the Society of Arts, and to the British Association on subjects ranging from the ginning of cotton and the manufacture of encaustic tiles to American railroad bridges. For his paper on railroad bridges, he was awarded the Telford Medal of the Institution of Civil Engineers.

On 5 January 1866, Colburn started yet another new weekly journal, this one entitled *Engineering.* In size and format, *Engineering* resembled closely *The Engineer.* On the other hand, Colburn's first editorial, "Breaking Ground," in the first number of *Engineering,* laid out the differences between the two journals that readers might expect. This journal is begun, he wrote, "because its conductor is convinced that there is no other worthy to represent Engineering." Asserting that nobody knew better than he "*how* so-called Engineering journals are conducted," he explained some of their shortcomings. They were seldom conducted by engineers; the "scissoring" of editors frequently took, without acknowledgment, accounts of new bridges, blast furnaces, and railway stations from *The Times,* "which really contains far more engineering *news* than all the technical papers taken together."[28]

Colburn was critical of patents in technical journals. New patents, he wrote, which could be cheaply collected at the Patent Office, were presented by editors of the technical papers (*The Engineer* was one of those papers) as "worthy engineering matter." Patents were granted, he admitted, to encourage useful inventions, but many were "trash" and many more were of little general interest. Furthermore, he wrote, most of those who ran the journals were also patent agents, who illustrated the patents of their own clients, "or of whoever will pay for engravings—not to say descriptions or commendations."

"*Engineering* may not reform these defects altogether," Colburn continued, "but it will do its best." Its articles, he promised, will be written by practical men. "It will describe and analyse, and if need be, criticize executed works to the exclusion of patented schemes, undigested by working." Of the thirty thousand words of text in the first fourteen-page number, Colburn wrote a very substantial part, probably well over one-half.[29] (Most scholarly articles today have fewer than ten thousand words.) And another number was due from the presses just seven days after the first was published.

Colburn was assisted by his friend William Maw, twenty-eight years old, a railway engineer, and James Dredge, twenty-six years old, a civil engineer and close friend of Alexander Holley. Maw helped with the reporting and writing, while Dredge made or procured the drawings that appeared in the journal. Both Maw and Dredge stayed with *Engineering* until they died, Dredge in 1906 and Maw in 1924.[30]

That first issue included detailed reviews of three important books, including Fairbairn's *Treatise on Iron Shipbuilding* (1865). That review included a critical discussion of fatigue in metal structures, a subject that still vexes designers and builders. An article on American railroads occupied more than a page—that is, over two thousand words; one on the Mont Cenis railway in France covered half a page, with three wood-engravings. Other articles—on Bessemer steel rails, deep-sea cables, the Ames gun, pumping machinery in Denmark, British blast furnaces, and the roof of St. Pancras railway station—were all intelligently reported. Such a diverse array of subjects and sites was to be typical of *Engineering*. In 1867, the Paris Exposition was covered extensively, probably by Colburn, who had made a particular effort to become acquainted with French engineering literature and practice.

In March 1866, less than three months after launching *Engineering*, Colburn boasted in his journal that he had on the books nearly a thousand permanent subscribers, including those most eminent in the profession in Great Britain, France, Germany, and America. There was, he wrote, no free list.[31]

Colburn's incredible industry and breadth of interests left its mark on the journals he edited. *The Engineer* and *Engineering* were and are the leading general engineering journals in the English-speaking countries of the world. Fifteen years after its commencement, *Engineering* began to be translated into French and published in Paris under the title *L'Ingénieur*,[32] but by that time Colburn was long gone from *Engineering*. As his friend Holley remarked of Colburn's experience with his short-lived American journal of 1860, "Although he learned to labor, he never learned to wait."[33]

Colburn quit *Engineering* in February 1870, proceeded to Paris and then to Boston. He avoided friends, wandered off to Belmont, and on 26 April 1870 killed himself. Maw and Dredge instantly picked up the pieces and proceeded along the same route they had traveled with Colburn. Except for the volume title page, which now identified Maw and Dredge rather than Colburn as "conductors," and the obituary notice of Colburn on 20 May 1870, no sign was evident of the change of editors.

Engineering yielded, over the years, an impressive series of books—most, of course, first published serially in the journal. The first, on the waterworks of London by Colburn and Maw, was published in 1867. Dredge visited the leading international expositions, reporting in the journal what he saw and publishing the 1873 Vienna and 1889 Paris expositions in book form. He was a British commissioner to the 1893 Chicago Columbian Exposition; his reports of the transportation exhibits in Chicago resulted in a large, impressively illustrated volume. With Maw, he published in 1872 a book of *Modern Examples of Road and Railway Bridges*. Other books by Dredge include a history of the Pennsylvania Railroad (1897), two volumes on electrical illumination (1882), a volume on modern French artillery (1892), and one on Thames bridges from the Tower of London to the source (1897).[34]

Colburn poured out his great talents into a variety of vessels, but his enthusiasm for engineering as a noble calling was finally embodied in those majestic volumes of *Engineering*. In papers and articles, the young American brought to British readers a picture of American as well as European engineering. After Colburn's death, Dredge made many visits to the United States and brought back exhaustive reports. Holley, a close friend of both Colburn and Dredge, contributed to *Engineering* a series of forty-one articles on iron and steel works in America.[35] Colburn may not have reformed all the defects of technical journals, but he did his best.

ENGINEERING NEWS

Engineering News-Record, now the leading journal in the field of construction of large civil engineering structures and systems, had its origin in Chicago more than a hundred years ago. On 15 April 1874, George Frost, thirty-five years old, a Canadian civil engineer and graduate of McGill College in Montreal, published the first issue of *The Engineer and Surveyor,* in sixteen two-column pages. The title was changed to *Engineering News* in 1875. The *News* was combined over forty years later, in 1917, with the New York journal *Engineering Record*.[36]

Frost had been a land surveyor with the Chicago and North Western Railroad when its first rails were laid west of Chicago. He justified his new journal by observing that civil engineering and land surveying were the only "learned professions" in America "unrepresented by a journal devoted exclusively to [their] interest."[37] He hoped eventually to raise his journal to "a standing in this country such as *Engineering* and *The Engineer* occupy in England."[38]

One of the preoccupations of Frost, which has survived to the present day in *Engineering News-Record,* was with the intellectual importance of failures. Structural and mechanical failures or design inadequacies were valuable opportunities for designers and constructors to learn what mistakes had led to the failures. A careful analysis of accidents, which Frost's journal sought to supply, might not only identify the mistakes but also provide insights into the performance of materials under unusual circumstances. "It has been truly said," Frost wrote in 1874, "that we learn as much from the failures of others, as from their success."[39]

Engineering News reported accidents and failures as they occurred. In January 1877, Frost observed that the December just past would be remembered as "replete with disasters almost unparalleled in this country, resulting in terrible loss of life." He listed the "Brooklyn holocaust" (a theater fire that killed nearly three hundred people), a shipwreck on Long Island that claimed thirty lives, and the "fatal plunge of the *Pacific Express* through the Ashtabula (Ohio) bridge" on Friday night, 29 December, in a blinding snowstorm.[40] In the last-named tragedy, an iron truss railroad bridge had collapsed under the load of two locomotives; all eleven passenger cars were dragged into the ravine below the bridge, where they immediately caught fire. At least eighty people perished. Frost published several pages describing and analyzing the accident, but his sources were second-hand newspaper accounts and reports of investigating committees.

The following year Frost was able to present first-hand reporting when by chance he was nearby when a construction accident occurred in New York City. On Friday, 14 June 1878, Frost was in the city on his way to a Boston meeting of the American Society of Civil Engineers. He heard in the streets of an accident to the cables of the Brooklyn Bridge, then under construction. Frost hurried to the foot of the cable anchorage abutment, where he found Colonel William Paine, engineer in charge of construction, and his chief assistant, Francis Collingwood, Jr. He also found there Park Benjamin, editor of *Scientific American,* William Wiley, son of the publisher John Wiley, and two reporters from the New York *Herald.* Paine invited the journalists to go up to the top of the anchorage, some eighty feet above the ground, to see what had happened.[41]

One of the four great cables that would suspend the bridge was being assembled. One of nineteen "strands" of wire that comprised the cable was being eased into place when the accident occurred. Each strand consisted of

278 individual wires, each about one-eighth of an inch in diameter. The strand, well over half a mile in length, was draped over the towers; the other end was anchored on the Brooklyn side while a winch on the New York cable anchorage pulled in or let out the end of each strand enough to match the length of the other eighteen strands. The single strand weighed some sixty thousand pounds. The winch line parted when the strand put a strain on it; the end of the strand whipped up over the New York tower and fell into the river, wetting some passengers on the ferryboat it barely missed. Two men were killed—one was knocked off the anchor platform to the ground; the other was crushed by the heavy strand as it took charge. Two other men were badly injured. Apparently a socket attaching the strand to the winch rope had failed, but the engineers ventured no guesses until they had had time to inspect the apparatus. The assistant engineer, Collingwood, prepared a report on the accident and presented it a few days later at one of the sessions of the civil engineers' meeting in Boston. Frost was thus able to report what he saw, as well as the results of his discussions with the responsible engineers.

In 1887, when Frost reflected on several more recent man-made disasters, he took the occasion to expand some of his earlier thoughts on failures. He wrote, "We could easily, if we had the facilities, publish the most interesting, the most instructive and the most valued engineering journal in the world, by devoting it to only one particular class of facts, the records of failures. . . . For the whole science of engineering, properly so called, has been built up from such records." Frost recognized the difficulty of obtaining the facts when failures occur, because the natural tendency of those in charge is to cover up the unpleasant events, hoping they will be quickly forgotten. His editorial ended with a plea to readers to tell the whole truth to the editors when they visited the scene of failure.[42] The *Journal of Failure* was never published, but *Engineering News* editors have made special efforts to visit the scene of an accident as quickly as possible after it occurred. Dam failures, roof collapses, and other structural failures have been reported regularly and matter-of-factly, often with the cooperation of those most closely involved.[43]

Later editors of *Engineering News* naturally had other items on their agenda besides analyzing failures. For several years after 1903, for example, the editors urged their readers to accept the notion that workmen's compensation laws, already "adopted by every industrial nation except the United States," were not only inevitable but also necessary, if the industrial world was to be fair to its workers.[44] Nevertheless, the push in the direction of failure analysis supplied by the first editor continues to the present and has resulted in an impressive number of valuable studies of events that otherwise might have occurred in vain.

RAILROAD GAZETTE

The proudest accomplishment of Matthias Forney, editor and owner of a number of railroad magazines from 1870 to 1896, was the campaign he waged in *Railroad Gazette* to reveal what he called the narrow-gauge fallacy. He was successful, he thought, in stemming a rising tide of enthusiasm for narrow-gauge railroads in the 1870s and bringing to their senses investors and others who believed promoters' promises of low costs and high profits.[45] Forney's efforts were not the only reason for the passing of the narrow-gauge mania, but constant pressure exerted over six years by a widely known editor of a leading journal surely dampened the ardor of enthusiasts as they learned some of the insurmountable problems presented by narrow-gauge railroads.

Forney's campaign began in 1870 when he became associate editor of the Chicago journal *Railroad Gazette*. Then thirty-five years old, Forney had worked for fifteen years as a draftsman and designer of locomotives in the shops of Ross Winans, in the Baltimore and Ohio shops, and in the Chicago shops of the Illinois Central Railroad. In Chicago, he had designed a tank-type locomotive, in which the water supply was carried on the locomotive frame rather than on a separate tender, thus increasing the adhesion of the driving wheels. The Forney locomotive, patented in 1866, was similar in principle to a British tank locomotive patented in the same year by Robert Fairlie, a thirty-five-year-old railroad superintendent in northern Ireland.[46] Fairlie put one of his locomotives on the Festiniog Railway, a narrow-gauge quarry line in northern Wales. He was so taken by the little railway that he soon became a publicist for narrow-gauge railways in general. The similarities between the Forney and Fairlie locomotives may or may not have had something to do with the vituperative tone of the controversy regarding narrow-gauge railroads.

In 1870, Fairlie read a paper at the annual meeting of the British Association for the Advancement of Science, setting forth arguments for building narrow-gauge rather than standard-gauge railroads.[47] A gauge of 4 feet 8½ inches between rails had been adopted when British railways began operating in 1825, and was adopted by the first American railroads. More than forty years' experience had shown that the gauge was reasonable, being wide enough to accommodate heavy rolling-stock, yet not so wide as to be cumbersome, as Brunel's seven-foot gauge on the Great Western Railway (1835) had proved to be. It was recognized by many, however, that the so-called standard gauge was an arbitrary width whose chief advantage was the fact that it was in common use, making it relatively easy to transfer rolling stock from one railway system to another.

Fairlie claimed, on the basis of a little sanguine analysis and almost no experience, that the cost of building a narrow-gauge railway was substantially less than that of standard-gauge and that locomotives and cars would also be less costly. Advocating gauges from two feet to three feet six inches, Fairlie claimed that the low cost of narrow-gauge railways would make possible a profitable operation on lines of light traffic. Furthermore, the lighter rails and locomotives would permit narrow-gauge railways to penetrate mountainous regions, another argument that at first glance seemed reasonable.

Forney reprinted Fairlie's paper in *Railroad Gazette* and proceeded to expose its fallacies. The issue was far from clear-cut, however. Forney argued that costs were not much different for standard- and narrow-gauge railroads and that if traffic would not support a lightly built standard-gauge system, neither would it support a narrow-gauge system. Yet Forney's reasons could not be readily confirmed because strictly comparable narrow- and standard-gauge systems did not exist. Then as now, somebody else's estimates were unlikely to sway a contractor or promoter who could find investors to whom Fairlie's arguments seemed plausible.[48]

The British journals *Engineering* and *The Engineer* had accepted Fairlie's gospel, as had many others throughout the world. For a time the construction of narrow-gauge systems became the rage. Few investors sought a second opinion, as had the proprietors of the proposed Texas Pacific Railroad, whose chief engineer advocated a narrow-gauge system some fifteen hundred miles in extent. The consultant, Silas Seymour, an experienced civil engineer, advised against the scheme, and it was dropped.[49] Best remembered of the narrow-gauge railroads is the Denver & Rio Grande. Within ten years of its commencement, sections were being converted to standard-gauge. Only parts of the line that were too remote and lightly traveled to justify rebuilding remained narrow gauge. Today, tourists note the quaintness of the "baby railroad" on the two sections of line not totally abandoned.[50]

The fatal disadvantage of narrow-gauge railroads was their inability to haul other lines' freight cars over their own tracks. Breaking bulk at transfer points was both expensive and unacceptable to those who preferred to have carload lots of merchandise go undisturbed to their destinations. Some railroads, such as the Denver & Rio Grande, laid a third rail on heavily traveled lines so that standard-gauge cars could be hauled, but that jury-rig was too cumbersome.

In 1877, Forney was able to report that *The Engineer* had changed its view of narrow-gauge railroads and that the journal had retracted "much if not all" of its former arguments for the system. Reflecting on the changing views (for which he took some credit), Forney thought Fairlie's paper of 1870 was "a remarkable example of ingenious sophistry" that appeared just at the time when it was becoming clear that "many railroads would not fulfill the expectations of those who built them."[51] The arguments—founded on hopes, not experience—were attractive to those seeking to escape an unpleasant situation. Angus Sinclair, writing thirty years later, recalled Fairlie as "able, earnest, indefatigable; he laid hold of and boldly advanced every argument which would serve his turn, good, bad, and indifferent, and he met with the success which determined enthusiasts generally achieve."[52]

In the United States, narrow-gauge mileage never exceeded about nine per cent of total railroad mileage and by 1914 had shrunk to less than two per cent. In 1915, the proportion of narrow-gauge mileage worldwide was about sixteen per cent. In India, after independence, mileage was evenly divided between broad gauge and narrow gauge: five feet, six inches, twenty thousand miles; meter gauge, sixteen thousand miles; gauges narrower than one meter, four thousand miles.[53] Perhaps the American figures reflect the influences that a determined, respected editor can have.

CONCLUSION

This brief foray into the editorial haunts of a few technical journals has illustrated the very different roles that different editors have played. The editors who worked for Munn & Co. were promoters of the myth that a patent—any patent—is a key to wealth. Zerah Colburn had a wide-ranging, critical mind and filled his columns with useful and interesting information. George Frost came to his journal with an obsession regarding engineering failures—an obsession that a hundred years later still insures full disclosure of errors and hazards when a structure or machine or system fails. Matthias Forney carried on a crusade against narrow-gauge railroads that deflated the extravagant claims of promoters and visionaries who saw possibilities unencumbered by realities. In order to use those journals intelligently as historical sources, we should know what was on an editor's agenda, how his ideology influenced the words we read, what hobby or obsession or loyalty may stand behind the campaigns and crusades that we encounter.

Another important consideration is the role of advertisers as well as the make-up of its readership itself. Although our foray yielded scant evidence regarding the supporters of technical journals, Zerah Colburn did drop a rare hint as to where the subscriptions to *Engineering* came from. Michael Borut told us how many subscribers *Scientific American* had, but his numbers come from Orson Munn's diaries, not from the journal itself. An appreciable part of a journal's income was derived from advertising, but because it was customary until recently to remove covers and advertising pages before binding journals in volumes, it is sometimes difficult to determine even the extent of advertising. As well as giving an indication of the support derived from advertising, the advertising pages sometimes tell us something about a journal's readers.

To peruse advertising in *Popular Mechanics,* for example, is to enter a world of self-improvement and marginal mail-order entrepreneurs. A typical number in 1930, selling for twenty-five cents, contained 175 pages of text and an additional 100 pages of advertising, including 17 pages of classified advertisements.[54] Numerous trade schools, including the International Correspondence School, offered courses by mail or in evening classes; books were advertised to improve one's voice, learn to be a salesman, play the piano, or become a magician; courses in manly muscular development involved books and apparatus; agents were sought to sell a bewildering array of gadgets and junk; a score and more of patent solicitors sought business; and a relatively few pages of advertising were devoted to real products, such as motorcycles, masonite, gasoline, and automobiles. Because many of the same advertisements persisted for years, we can assume that readers responded often enough to make advertising profitable. An editor's approach to his readers would inevitably be affected by his perceptions of the kinds of readers who would patronize his advertisers, and thus another skewing force would be present to resist or accommodate.

The motives and purposes of editors (and publishers, when an editor was not also publisher) were varied and full of subtleties, but we can be sure that few editors saw their calling as merely a job to be done in order to collect a weekly pay envelope. Some nineteenth-century editors and publishers would, I am sure, have something in common with Neil Ellis, owner of the Manchester (Connecticut) *Journal Inquirer,* a daily tabloid newspaper. In December 1986, when Ellis was arrested for bribery and conspiracy, his editor carried the story "straight" because he had "no instructions from the owner" as to how to write up or display the incident. Commenting on Ellis and his *Journal Inquirer,* the editor told a *New York Times* reporter: "He didn't start this newspaper as a money-maker. He started it because he had a certain view of the world and wanted to raise hell."[55]

NOTES

I wish to acknowledge Mel Kranzberg's timely help and encouragement when this engineer was becoming a historian. In writing this essay, I am particularly indebted to Charles E. Haines for ideas, discussions, encouragement, and several key sentences.

1. Leonard Reich, "Industrial Research and the Pursuit of Corporate Security: The Early Years of Bell Labs," *Business History Review* 54 (1980): 504–29, esp. 515.

2. Eugene S. Ferguson, "John Ericsson and the Age of Caloric," *U.S. National Museum Bulletin* 228 (1963): 41–60.

3. Michael Borut, "The *Scientific American* in Nineteenth Century America" (Ph. D. dissertation, New York University, 1977). I am indebted to Borut for most of the facts not derivable from the journal itself. Borut has shown how misleading are the official histories published in the journal.

4. Ibid., pp. 42–48.

5. Ibid., pp. 85–88.

6. Ibid., p. 89.

7. Circulation rose from eight thousand in 1848 to twenty-five thousand in 1859 (ibid., pp. 67–69); in vol. 4 (1848–49), two hundred ninety-nine pieces were copied from one hundred forty-four American newspapers and other periodicals, forty-seven foreign sources, and twenty-nine books and society proceedings (ibid., p. 74).

8. See, e.g., "Advice to Inventors," an advertisement in *Scientific American* 15 (3 March 1860): 155–56. Quotation from Munn's diary, 17 February 1852, in Borut, "*Scientific American* in Nineteenth Century America," p. 150.

9. *The Scientific American Reference Book* (New York: Munn & Co., 1876), p. 3.

10. Robert C. Post, *Physics, Patents, and Politics. A Biography of Charles Grafton Page* (New York: Science History Publications, 1976), pp. 49–51, 107, 110–15. Post's book introduced me to the "patent lobby"; see his index for relations of *Scientific American* with the U. S. Patent Office.

11. *The Scientific American Reference Book,* pp. 32–33.

12. Borut, "*Scientific American* in Nineteenth Century America," pp. 230–35.

13. *Scientific American,* n.s. 19 (1 July 1868): 14; 96 (9 March 1907): 207.

14. Borut, "*Scientific American* in Nineteenth Century America," pp. 90–91; Henry T. Brown's *Five Hundred and Seven Mechanical Movements* (New York, Brown and Seward, 1896), includes an advertisement for his firm, Brown & Seward.

15. I saw *American Inventor* some years ago in an overflow serial room in the National Museum of American History. My notes have excerpts only from vols. 1 (1878) and 4 (1881).

16. *Scientific American* 15 (3 March 1860): 155–56. U.S. Bureau of the Census, *Historical Statistics of the United States, Colonial Times to 1957* (Washington, D.C.: U.S. Government Printing Office, 1960), p. 14, gives a total of 594,515 souls in towns of 2500–4999 population. I should guess the number of towns at two hundred.

17. Post, *Physics, Patents, and Politics,* p. 160, says "there were nearly three dozen agencies in Washington alone, and at least twice that many throughout the rest of the country. . . . [Munn & Co. handled] more than the next 15 firms combined." Only a few agents handled "even a two or three per cent share of the market."

18. *Scientific American* 74 (11 January 1896): 18–19. This is the obituary notice of Beach.

19. Carroll W. Pursell, Jr., "Testing a Carriage: The 'American Industry' Series of *Scientific American,*" *Technology and Culture* 17 (January 1976): 82–92; David A. Hounshell, "Public Relations or Public Understanding? The American Industry Series in *Scientific American,*" *Technology and Culture* 21 (October 1980):589–93.

20. Borut, "*Scientific American* in Nineteenth Century America," pp. 261–68.

21. Since reading Thomas P. Hughes's *Elmer Sperry, Inventor and Engineer* (Baltimore: Johns Hopkins University Press, 1971), I have carried in my mind a picture of young Sperry, in his teens, in an upstate New York YMCA reading room, devouring such journals as *Scientific American.* My recollection was not precise (see Hughes, p. 15), but that passage sharpened my appreciation of the influence of such institutions. Michael Pupin, in his autobiography, told of reading *Scientific American,* c. 1875, "assiduously with the aid of a pocket dictionary." See Michael Pupin, *From Immigrant to Inventor* (1923; reprint, New York: Scribners, 1960), p. 74; quoted in Borut, "*Scientific American* in Nineteenth Century America," pp. 284–85.

22. *The United States Patent Law. Instructions How to Obtain Letters Patent* (New York: Munn & Co., 1871), p. 100. This is another of the pocket-size pamphlets issued to encourage patenting.

23. The recognition that *Scientific American* charged for everything came to me slowly. Hounshell, "Public Relations," discovered in the McCormick papers a fee of five hundred dollars for an "American Industry" installment. Close reading of many advertisements in the journal finally convinced me that whenever praise was given in the text it had been paid for by the one being praised. *Engineering News* 1 (15 October 1874): 135; *Railroad Gazette* 9 (24 August 1877): 386; *American Machinist* 4 (1 January 1881): 8.

24. Edward L. Throm, *Fifty Years of Popular Mechanics 1902–1952* (New York: Simon and Shuster, 1952), p. ix.

25. Ibid., pp. 216–18; *Popular Mechanics* 54 (August 1930): 214–19.

26. *The Engineer Centenary Number 1856–1956* (London: Morgan Bros. Ltd., 1956), p. 146.

27. Biographical details are from obituary notices in *New York Times* (2 May 1870) (presumably written by Alexander Holley), and in *Engineering* 9 (20 May 1870): 361. Titles of books published by Colburn are from the *Engineering* obituary and library catalogs.

28. *Engineering* 1 (5 January 1866): 1–2.

29. This statement is deduced from a close reading of the journal; it is, of course, a guess.

30. An obituary notice of Dredge appeared in *Engineering* 82 (24 August 1906): 241–42; a biographical sketch is in *Dictionary of National Biography,* suppl. 2 (Oxford: Oxford University Press, 1901–11); Maw's obituary notice was in *Engineering* 117 (21 March 1924): 371–74.

31. *Engineering* 1 (23 March 1866): 185.

32. Henry C. Bolton, *A Catalogue of Scientific and Technical Periodicals, 1665–1895,* 2nd ed., Smithsonian Misc. Colls. vol. 40 (Washington, D.C.: Smithsonian Institution, 1897), entry no 2184.

33. In the *New York Times* obituary notice (n. 27 above).

34. See Dredge's biography (n. 30 above).

35. Titles and dates of articles are in *Memorial of Alexander Lyman Holley, C.E., LL.D., 1832–1882* (New York: American Institute of Mining Engineers, 1884), pp. 146–48.

36. An obituary notice of Frost (1838–1917) is in *New York Times,* 16 March 1917. A ceremonial but useful history of the constituent journals is in "History by the Week," *Engineering News-Record* 143 (1 September 1949): A24–32.

37. *Engineering News* 1 (15 May 1874): 17.

38. *Engineering News* 1 (15 April 1874): 8.

39. *Engineering News* 1 (15 July 1874): 62.

40. *Engineering News* 4 (6 January 1877): 7.

41. This and the following paragraph are based on *Engineering News* 5 (27 June 1878): 201; 5 (22 August 1878): 265–66. The accident is also described in David McCullough, *The Great Bridge* (New York: Simon and Shuster, 1972), pp. 350, 438–40.

42. *Engineering News* 17 (9 April 1887): 229, 237–38.

43. The latest book compiled from articles published in *Engineering News-Record* is Steven S. Ross, *Construction Disasters: Design Failures, Causes, and Prevention* (New York: McGraw-Hill, 1984).

44. Quotation is in *Engineering News* 59 (16 April 1908): 425.

45. "Matthias Nace Forney: A Biographical Sketch," *Cassier's Magazine* 33 (March 1908): 594–96.

46. *Dictionary of American Biography* (New York: Scribners, 1927–36), s.v. Forney, Matthias Nace (1835–1908); *Engineering* 40 (7 August 1885): 133.

47. *Railroad Gazette* 9 (24 August 1877): 386.

48. Ibid.

49. The sense of Seymour's report to the Texas Pacific promoters can be found in Robert F. Fairlie, *Railways or No Railways. Narrow Gauge, Economy and Efficiency, v. Broad Gauge, Costliness with Extravagance* (London: E. Wilson, 1872).

50. John H. White, "The Narrow Gauge Fallacy," *Railroad History* 141 (1979): 77–96.

51. *Railroad Gazette* 9 (24 August 1877): 386.

52. Angus Sinclair, *Development of the Locomotive Engine* [1907], annotated ed. John H. White (Cambridge, Mass.: MIT Press, 1970), p. 193.

53. White, "The Narrow Gauge Fallacy," pp. 84, 92; J. N. Westwood, *Railways of India* (Newton Abbot: David and Charles, 1974), p. 42.

54. *Popular Mechanics* 54 (August 1930).

55. *New York Times,* 4 December 1986, p. B7.

Technology on Its Toes: Late Victorian Ballets, Pageants, and Industrial Exhibitions

BRUCE SINCLAIR

Historians of technology know a great deal about the nineteenth century's remarkable inventiveness, and something about the impact of those changed ways of doing things, but very little of how ordinary people viewed those changes. That is partly a matter of sources; the materials we usually study do not tell us much about social context. Yet our difficulty is also a matter of focus, and if we mean to claim a relation between technology and culture we need to get beyond simple notions of cause and effect.

Although the apprehension of popular sentiment is inherently difficult, the sources for such an inquiry are not so hard to find. One of the nineteenth century's most often staged ballets, for instance, was a triumphant celebration of technological progress called *Excelsior*. In the entire history of dance few productions have enjoyed such success, and, because the ballet's creator aimed at a mass audience, *Excelsior* provides a ready-made opportunity to examine the relations between technology and its cultural setting.

The ballet proved instantly popular. Following its Milan premiere at La Scala in March 1881, it was repeated more than a hundred times that year, and subsequently performed there thirty-one times in 1883, forty-five times in 1888, and twenty-seven times in 1894. It also proved the rage in Paris, London, Berlin, St. Petersburg, and particularly in Vienna, where it remained in the Court and State Opera Company's repertory continuously from 1881 through 1910, for a total of 329 performances. Nor in fact did the ballet's celebrity end there. It was revived in Florence in 1967 and then, ninety-four years after its premiere, performed again at La Scala in 1975.[1]

Luigi Manzotti, who choreographed *Excelsior* (and who gained wealth and fame from its popularity), did not call it a ballet. Instead, he described it as an "Azione Choreographica, storica, allegorica, fantastica"—a combination that in common with ballet at the time involved both dance and mime. Choreographed for a huge cast of about five hundred dancers, *Excelsior* was meant to be a grand-scale spectacle, tracing the progress of civilization through

71

Cover illustration from the program of the 1881 premiere production of *Excelsior*.

Scene from the premiere performance of *Excelsior* at LaScala. (*From* Il Teatro Illustrato, *March 1881.*)

technical advance in a series of historical episodes that symbolized the
eventual victory of reason over superstition and ignorance.

Manzotti's earlier ballets, one about Michaelangelo, for instance, and
another about Galileo, concerned themselves with the ideals of the Risorgi-
mento, mixing pride in the past accomplishment of Italians with patriotic
convictions of the nation's future. *Excelsior* brought the vision of technological progress to those themes and linked Italy's destiny to a new era of
international brotherhood through applied science. Cyril Beaumont, in his
Complete Book of Ballets, says that notion came to Manzotti while in Lyons
when he became aware of the extensive industrial activity of that city. In the
introduction to the libretto, however, Manzotti claimed it was seeing the
Italian entrance to the Mont Cenis tunnel through the Alps that inspired
Excelsior. The monument there suggested to him the dramatic contrast
between the obscurantism of the Spanish Inquisition and enlightened rationality epitomized by the engineering feat that opened easy communication
between Italy and France.

Excelsior begins with just such a stark delineation. The first scene shows a
Spanish town, destroyed by the evil force of religious intolerance, which is
exemplified in the sound of a bell tolling for those about to be burned at the
stake. An allegorical figure, the spirit of Darkness, presides over this dismal
display, gloating in his victory as another allegorical figure, a beautiful young
woman, lies at his feet in chains, amidst a litter of instruments of torture. She
is the spirit of Light, obviously the antithesis of all that Darkness represents.
As lightning flashes around her, she gradually revives to break her chains,
telling the spirit of Darkness that his time of triumph has ended and that she
will command the future.

To demonstrate her prophecy, she leads us at the opening of the second
scene to the Palace of Genius and Science, whose walls are inscribed with the
names of those made famous throughout the ages for their accomplishment
and whose work forms the scientific treasure of the modern era—steam
power, the electric telegraph, the Suez Canal, and the Mont Cenis tunnel.
The palace is inhabited by a number of other allegorical characters—Science,
Power, Industry, Love, Civilization, Perseverance, Union, Concord, Courage,
Glory, Invention, the Fine Arts, Agriculture, and Commerce—who in small
groups do a military parade-like set of dances by way of paying homage to the
principal allegorical figure, Civilization, who is the prima ballerina. She and
the spirit of Light meet in triumph, to end part I of *Excelsior.*

The ballet then shifts to a series of quasi-historical scenes that comprise
the next four parts. The initial scene has to do with the invention of the
steamboat. Edwin Evans, who had attended the first London production in
1885 when he was eleven years old, and who still remembered it in 1942
better than many ballets he was later to see, recalled that scene with particular fondness. In it, Denis Papin, a Huguenot exiled from France by the

revocation of the Edict of Nantes, had taken his steamboat to the Weser River, near Bremen, for its preliminary trial. The action begins in a small village on the banks of the river, where the local tavernkeeper's son, Valentine, has just come home victorious from a regatta of boatmen, to the congratulations of his parents, his fiancée Fanny, and his friends. But the defeated opponents still feel jealous of Valentine's success (emotions that provide the occasion for a mazurka to calm everyone down), and new races are planned. However, as the boatmen go back to the river, the spirit of Darkness shows them Papin's steam-powered vessel, which runs without the need for human energy. Goaded on by that evil genie, who tells the boatmen that Papin's new invention will ruin their trade, they destroy it—just as people with Manzotti's message liked to believe had actually happened.

Light suddenly appears on the scene then, not only to save Papin's life, but also to rebuke the spirit of Darkness with the proclamation: "Here, Enemy, Look at my future Works!" With that, the stage is marvelously transformed into a "grandiose *tableau vivant* of Steam Power in all its Glory," which shows the skyline of New York City and the Brooklyn Bridge, with steam-powered trains crossing the bridge while steamboats ply the river beneath.[2] The Brooklyn Bridge was actually still under construction in 1881, and in any event Manzotti did not know what it was supposed to look like, so the drawings of it on the cover of the ballet's program do not bear any resemblance to the finished product. Nonetheless, it was "awfully exciting," Edwin Evans said, and many of those who attended the 1975 La Scala revival felt the same.

Part III also has two scenes, the first of which depends again on historical allusion. The setting is Alessandro Volta's laboratory at Como, where he is shown trying to produce an electric spark from his newly developed battery while Darkness, in the role of kibitzer, tries to create doubt in the scientist's mind. Volta's genius finally prevails over these difficulties, however, and after a brilliant electric spark illuminates the stage, he runs to tell others of his success. The spirit of Darkness seizes the opportunity of Volta's absence to go over to the battery intending to destroy it. But he is ignorant of its powerful force and gets a nasty shock that sends him scuttling off stage. Then, as she had done for Papin, Light reveals to Volta the future importance of his invention, and the curtain falls to the music of electric bells and sparks of electricity in a grand divertissement. In a second scene, set in Telegraph Square in Washington, D.C., Light glories in this victory of applied science. She is surrounded by swarms of little telegraph messengers, and they are decorated with small light bulbs that they switch on and off as they dance. Toward the side of the stage, however, the spirit of Darkness swears revenge.

Part IV of *Excelsior* follows the same format; one scene demonstrates the evil power of the spirit of Darkness, while the ensuing scene portrays the subsequent victory of progressive and enlightened technology. This time, the

Illustration from the premiere performance of *Excelsior.* (*From* Il Teatro Illustrato, *March 1881.*)

example contrasts the hazards of desert transportation before the construction of the Suez Canal with the easy flow of prosperous traffic that followed, as it were, in its wake.

The curtain rises on a desert setting, with a camel caravan slowly making its way through the sands. Every part of Manzotti's ballet emphasized the spectacular, and this scene was no different; real camels trudged across the stage. But then a dreaded *simoon,* one of those sudden, hot, suffocating sandstorms (a little less convincingly suggested by great swirling veils of light colored cloth), suddenly enveloped the caravan, plunging it into dismay and disorder. Always ready to grasp such advantage, the Prince of Darkness conjures up a band of desert pirates—in the premiere performance riding matched Arabian horses that Michele Strogoff had brought from Paris for an international equestrian exposition. These ruthless marauders pillage the disabled caravan, and the unfortunate wayfarers, unable to protect themselves against either the brigands or the storm, are gradually buried in the sand.

According to one reviewer, that scene, for what must have been a variety of reasons, left the audience "in a state of shock."[3] But as she had done before, Light points to the horizon to reveal her ultimate triumph, and the subsequent change of curtains shows a broad and calm waterway, bordered by verdant banks, stretching through the desert. That beautiful panorama, included Ismailia, the so-called Venice of the Desert, according to Manzotti, because of its gaiety, its vitality, its gardens, and its little palaces. Marvelously decorated triumphal arches celebrate DeLesseps's engineering accomplishment in a festival marked by dancers in the native costumes of Egypt, the Sudan, and Ethiopia.

The last of *Excelsior's* historical episodes involves digging the tunnel through the Alps at Mont Cenis, what Manzotti called "L'Ultima Mina." In this final ordeal, the Italian miners are shown working fiercely to break through to the French side, interrupting their labors only to listen for the sounds of their approaching counterparts. When they fail to hear those encouraging noises, the spirit of Darkness plays on their consternation to bring a halt to the work. But the chief engineer then catches the sound of the French crew and inspires his men to attack with renewed energy, as Light comes to their aid by bringing down the final barrier. In a theater-shaking crash, French miners emerge from billowing plumes of dust, and the stage is filled with begrimed workers, rejoicing and embracing one another. At that point, the nemesis of progress tries to sneak away, to pursue his nefarious game another time, but Light apprehends him, telling him that his end has now come. As clouds float over the stage, one can see through them that all the nations are at peace. At that point, with a signal from Light, the earth opens up to swallow forever the tyrannical ignorance Darkness represented.

In the final part of the ballet, the apotheosis, all the nations of the world are gathered in the Palace of Civilization, each one represented by a cohort of five dancers in national colors and carrying huge flags. Then comes a *pas de deux* danced by Civilization and Mankind, who are subsequently joined by the rest of the troupe. Parading and marching continue "until the whole stage is a sea of uniforms and flags and galoping, and kicking ballerinas and premier danceurs, corps-de-ballet dancers of all grades, pupils and supers." As a critic said of the 1975 revival, "it has to be seen to be believed!"[4]

Part of *Excelsior*'s enormous popularity flowed directly from its spectacular effects and from the sheer numbers of dancers. The precision required to coordinate the movements of that many people on stage was so far beyond anything previously produced that military metaphors inevitably came to mind. When the ballet was presented in Paris, for instance, one critic remarked that the staging might well have been the work of the German General Staff, and that at any moment during the performance he almost expected to see Field Marshall von Moltke enter on horseback.

In fact, the reigning style of Italian ballet, which *Excelsior* exemplified, emphasized exacting choreography, and another metaphor that emerged from the analysis of this novel approach connected it to mechanism. "Never before have I seen an automaton perform with such mathematical precision," a commentator wrote of Elena Cornalba, one of the three prima ballerinas of *Excelsior*'s nine-month run in Paris. "Cornalba," he went on, "performs these steps with the precision of a chronometer." Then, in an image his audience could easily grasp, the critic likened her "legs of iron" and mechanical regularity to the action of those sewing machines that industrial exhibitions had accustomed people to seeing.[5]

That connection suggests another reason for *Excelsior*'s success. In place of the typical themes of romantic ballet, Manzotti's production was one of the earliest attempts to capitalize on subjects of current popular interest—in this case, the dramatic changes brought about by science and engineering. One can find a few prior examples. Ippolito Monplaisir choreographed a ballet called *L'Almea* in 1872 that was inspired by the opening of the Suez Canal. But it was really *Excelsior* that stimulated such interest in a new subject matter for dance, a shift that coincided with the emergence of a different sort of theater and audience.

The Eden theater in Paris, for instance, which opened with *Excelsior* in 1883, was designed more as an amusement house, aiming to combine circus and Italian ballet—not inappropriately for a production that brought live animals on stage. The auditorium was decorated in what one journalist called "the Egypto-Assyrio-Indian style"; with various bars in the theater's hallways each reflecting a national culture—Spanish, Russian, American, Swiss, Dutch—with barmaids in the traditional costume of the country, a concept introduced at the Paris Universal Exposition of 1867. One critic's analysis of the Eden's choreography implies something of the audience, too. The dancing was, he said, "crude, nude, brutal, democratic—an apotheosis of femininity placed within reach of all; for the ballerinas of the Eden dance without *tutus*,

Virginia Zucchi in the prima ballerina role of "Civilization" in Manzotti's *Excelsior.*
(Photo courtesy of Marcel Dekker, Inc.)

and their beauty is revealed brutally without the intervention of the light cloud of gauze that would render it more desirable."[6]

In England, too, dance's ethereal nature was transformed by the demand for novelty. Topical, or, as it was called there, "up-to-date," ballet, fitted easily into music hall traditions. Two theaters in particular, the Alhambra and the Empire, aimed to popularize ballet, and their productions, Manzotti-like, were also characterized by grand scale and lavish effects. The Alhambra, in a connection that again suggests links between technology and amusement, had originally opened in 1854 as the Panopticon of the Arts and Sciences, but by 1900 it was staging spectacles such as *Soldiers of the Queen,* which aimed to capitalize on sentiments aroused by the Boer War. Another of Alhambra's patriotic ballets was *Brittania's Realm,* with dances by Carlo Coppi who had staged a version of *Excelsior* there. It was an equally sumptuous production, and among its charming novelties was the *"Pas des Patineurs"* in the Canadian Skating Carnival scene—an image of that country exactly like the one presented by the displays of furs and sleighs that were so often a feature of Canadian exhibits at the international exhibitions.[7] The Empire also emphasized spectacle and topicality by using electric lighting in novel ways. So, for example, an 1889 production entitled *The Paris Exhibition* utilized luminous fountains on its set and an Eiffel Tower that, just as the one in Paris, lighted up at dusk.[8]

All those techniques that the Eden, Alhambra, and Empire theaters employed so successfully—light, color, novelty, scale, and action—were, of course, the outstanding ingredients of the late nineteenth-century's industrial expositions, even more calculated forms of popular technological entertainment. Indeed, one does not have to search for parallels; in Manzotti's case, Italian critics explicitly connected his ballet to "Exhibitionmania."[9] But besides shared production values, *Excelsior* and the expositions dealt in the same symbols and infused them with the same messages. So, for example, the most popular exhibit at Paris in 1867, and the one that best captured the spirit of nineteenth-century progress, was a large-scale model of the Suez Canal, just opened that year, where fairgoers could see ships passing through the waterway almost as in actuality. Similarly, at the Vienna Exhibition of 1873, the Italian government displayed its latest engineering triumph in an enormous model of the Italian entrance to the Mont Cenis tunnel. Complete with railway systems, signaling devices, and a full-sized train, it proved that year's most persuasive expression of engineering in the service of civilization's advance.

For people attending these celebrations of industrial achievement, just as for Manzotti, Suez and the Mont Cenis tunnel became emblems of mastery over nature and its obstacles. As a consequence, the choreographer could employ these icons in his ballet confident of their evocative power, and those who organized expositions could do the same. Eugene de Vogue's analysis of the Eiffel Tower shows how easily technological masterpieces could be made to speak to the largest issues:

> Soon we shall study a new form of design which one sees dawning as he inspects the Galerie des Machines; but the tower is a witness to its arrival. It symbolizes as well another dominant characteristic of the Exposition: the search for everything that might make communication easier and accelerate exchanges between and fusion of the races.[10]

The idea that engineering promoted international harmony by creating better communication among the peoples of the world was not novel to those fairs at Paris or Vienna, any more than it had been to Manzotti. But successive exhibitions after London's Crystal Palace elaborated and refined Prince Albert's vision of peaceful contests of industry into a powerful advertisement for European technology and those whose interests it served. The Paris Universal Expositions best embodied that development, and the person most responsible for articulating the ideology that linked technical change and the advance of Western civilization was the brilliant French engineer Frederic LePlay. Patrick Geddes, a close student of the nineteenth century's industrial fairs, described LePlay as a man of "personal and concrete familiarity with industry in all its forms [and] of mechanism and engineering characteristic of the century, but also of the deepest economic insight, of the freshest originality of social outlook, of the widest international relations."[11] Those were the talents that led Louis Napoleon to appoint LePlay director of the 1855 exposition and, as Paris became the acknowledged center of the international exhibition movement, he was able to influence their style and message until the end of the nineteenth century.

A social activist himself, LePlay's overriding concerns were didactic. He developed a classification system for exhibits, for instance, that had as its objective the quick and direct comparison of similar kinds of manufactured objects from different countries. The building he devised for the 1867 exposition, for which Gustave Eiffel did the structural engineering, was to serve that purpose, too. A gigantic iron and glass oval called the Palace of Industry, LePlay designed it as a series of six concentric galleries, with goods in the same category grouped along each ring of the oval. Besides encouraging the comparison of specimens within categories, the overall arrangement expressed LePlay's idea of the relation between technology and culture; he placed machine tools around the outer rings, manufactured products on the inner rings, and concluded with a display of the history of human advancement at the center. One moved from machines to modern civilization. To help the working class understand its role in this process, LePlay organized exhibits on the history of labor and arranged special tours for workingmen.[12]

Even the buildings themselves were meant to teach. LePlay emphasized what came to be called engineering architecture, in which the structure

Architecture as education. The Gallerie des Machines at the 1889 Paris Exhibition. (*From the* Illustrated London News, *23 March 1889, p. 356.*)

became an exhibit as well. That approach was perfectly exemplified in the great Gallerie des Machines in 1889—"a sanctuary of industry," as one critic said—and it was just through such exercises in 1867 and 1878 that Eiffel gained the experience for his tower. To cap the concept of exposition as educational experience, the engineer directors of the Paris fairs helped organize and provided support for international congresses on subjects such as weights and measures, statistics, and copyright legislation.

Although moving machinery had been exhibited before, LePlay made it an essential element of his educational program. In addition to individual machines turning out their separate products, he also gathered machinery to show sequential stages of manufacturing, and yet other displays of equipment to demonstrate particular technical processes. In that fashion, machinery—and especially the prime movers that powered it—often became the central symbolic elements of exhibitions, as did the great Corliss steam engine at the U.S. Centennial Exposition in Philadelphia in 1876, and the huge electric dynamos at Paris in 1900 that caused Henry Adams such distress. Most visitors, indeed, got the impression that moving machinery made the place, as they so often put it, "come alive."

To demonstrate even further the connection between technical advance and human progress, LePlay stressed novelty, emphasizing the inaugural display of new materials, such as aluminum, or new processes, such as the

"Etat Actuel." Building the Eiffel Tower for the 1889 Paris Exposition. (*From* L'Exposition de Paris, 1889, *vol. 1, p. 5.*)

distillation of petroleum. To give the concept of civilized progress through industry the sheen of internationalism, it was LePlay who first introduced the idea of ethnic pavilions, where visitors might see people from other countries, dressed in their native costumes, and where one might also taste foreign foods and hear foreign music.

LePlay's strict sense of purpose ruled out such naive displays of ingenuity as the model of Tintern Abbey made from cork by a retired clergyman that had found a place in the Crystal Palace. Indeed, an English commentator, writing a series of articles for the *Daily Telegraph* about the Paris Exposition, contrasted the innocence of the Crystal Palace exhibits with "the gigantic Bazaar" of 1878, where "individual man, save in a very few instances, disappears, and is replaced by great Companies and great Firms solicitous of orders."[13] As these periodic assemblies gained in commercial importance and popular appeal, the sideshows on the periphery flourished also, to LePlay's regret.

Although LePlay and his successor, Alfred Picard, another engineer graduate of the École Polytechnique, emphasized pedagogy, there was an ideology behind the lessons. These men were, after all, the same kind of engineers that planned and built the Suez Canal, that dug the Mont Cenis Tunnel. They attended the same engineering schools and looked for the same career successes. Committed to the application of their knowledge for the sake of national prestige, industrial might, and the winning of a colonial empire, they made the exhibitions serve that viewpoint, too. Their displays argued the brotherhood of man but demonstrated the power of engineering and commerce.

It is a commonplace among present-day historians of technology that their subject is an artifact of social, political, and economic forces. Mel Kranzberg meant to suggest something of that relationship when he named his journal *Technology and Culture*. The analysis of theatrical productions like *Excelsior* helps us to see the richness and complexity of these connections, and also their shifting nature. So by 1900, as exhibitions lost their earlier character, dance changed, too. The iconographic symbols of the Paris exposition that year—most noticeably the colossal statuary groups of rampant stallions, snarling lions, and muscular women—echoed older ideas in such titles as "Harmony Destroying Discord," or "Nature Disrobing Before Science." But the architectural styles were no longer inspired by engineering. They reflected instead a Beaux Arts eclecticism, ornately plastered structures that would scarcely last until the Exposition closed in early fall.

Dance at the turn of the century also seemed to be moving in a different direction. Instead of a cast of hundreds, and far from the paradelike, mechanical qualities of Manzotti's choreography, a young American woman named Loie Fuller created astonishing effects in a theater of her own at the 1900 exhibition, by dancing in the light of electric lamps, using different colored filters and flowing veils to suggest a simple, natural, uncorseted approach to life, spiritually as well as actually. And when compared with the *ballets mechaniques* of the 1920s, or with plays of the same era like Eugene O'Neill's *Dynamo,* in which engineers became faceless characters and their work was made sinister, *Excelsior* seems extravagant and far-fetched in its message.

Yet, paradoxical as it seems, Manzotti consciously aimed to break with older escapist themes and connect his art to modern issues. As explicitly as LePlay, Manzotti meant to instruct, and the critics recognized it. "The ballet-master," one of them wrote, "is not content nowadays to be artistic; he must be didactic as well."[14] Indeed, a writer in *Il Teatro Illustrato* called *Excelsior* "a philosophic ballet."[15] Inspired by the idea that technical change was progressive in its nature, Manzotti brought dance, music, costume design, and scenic art to the service of engineering and industry in the name of internationalism.

Art played a somewhat similar role for men like LePlay and Picard. Mid-century critics of modernism damned factories and railways on aesthetic and spiritual grounds alike. Locomotives spoiled the landscape, industry debased its workers, and mechanization led to shoddy goods. As a response to that critique, those who organized industrial exhibitions, from 1851 onwards, tried to integrate art and manufacture. The most obvious way was through industrial design—the explicit adaptation of art to manufacturing—and that became a staple concern of exhibition juries. But it was part of the same agenda generally to link culture and engineering, to put galleries of art next to galleries of machines, as complementary elements in the march of Western civilization.

For all his idealism, however, the effect of Manzotti's approach was to commercialize dance. What he invented was the "propaganda" ballet, and there is a clear line of descent from *Excelsior* to the patriotic productions of the Empire and Alhambra, designed to unify popular sentiment at the time of the Boer War.[16] Dance was thus transformed from a medium of poetic expression into spectacle, depending for its effects on scale and novelty. This mode of discourse also dealt best in absolutes, pitting science against superstition, dividing history into the dismal period before the invention of the steam engine and the modern era's brilliant achievements.

The Paris exhibitions spoke the same language. In a vocabulary that stressed size, energy, speed, and change, they presented their onlookers with the contrast between the industrial might of Europe and the exotic primitiveness of its imperial outposts—by day the colossal Gallerie des Machines, the "Palace of Power," by night Javanese dancing girls.[17] Experiencing these kinds of opposites stimulated a pleasing sense of cosmopolitanism that, among other things, helped legitimate overseas adventures, whose success depended precisely on the technological mastery exhibitions celebrated.

At the simplest level of understanding, nothing could seem more real than Eiffel's tower. But the consequence of presenting elementary ideas in the form of spectacular effect and in a context of persistent novelty was to blur the distinction between reality and fantasy. So, in the 1900 exposition's "mareorama," visitors could be made to feel through clever simulation that they were actually on an ocean voyage, or in another similar exhibit, traveling on a train across Russia. These fanciful forms of travel made people think of an imaginary future brought about by technological advance, when they might magically be whisked to faraway places, and led them naturally to link *Excelsior,* exhibitions, and Jules Verne.[18]

Displays of machines carried the same message: tomorrow would be even more marvellous than today. Rational enterprise created abundance and amity. So, whether in the sideshows or the main arena, a blitzkreig of sensations drew one's attention away from the contradictions inherent, even explicit, in these theatrical productions. Ceremonially, the 1889 Paris exposition was dedicated to the peaceful competition of industry, but the children's toys of the French exhibit were dominated by "military playthings"—gunboats, artillery and "tiny soldiers of every grade, with every appliance of modern engineering."[19] As historians, we know the nineteenth century's industrial exhibitions are important indicators of technical activity, and we know they served the institutional development of engineering during that era, too. But here, we see them in their guise as theater, artfully presenting a commercial message on behalf of a certain view of the past and a set of expectations about the future.

Although engineers seldom appear on the ballet stage, *Excelsior* can be situated in a broad cultural context. As a novel source for the history of technology, we learn interesting bits of information. For instance, Manzotti imagined Europe as the locus of inventive genius but viewed the New World as the place where steam power and electric energy would be fully realized. For our purposes, however, the utility of the ballet lies in its capacity to extend and enliven our understanding of the connections between nineteenth-century technology and an emerging middle-class culture. Those relations obviously go beyond mass production and mass consumption; machine images even permeated the era's choreography. Or, to put the matter the other way around, the better we can understand the nineteenth-century power of icons like the Suez Canal and the Mont Cenis tunnel, the better able we will be to explain the persistence of that ideology into the twentieth century. Among aesthetes, *Excelsior* never enjoyed great favor; Tchaikovsky thought it unbelievably stupid.[20] The important impact, however, and the one with which we should be concerned, was on a different audience. Furthermore, we can see these ideological effects lasting into our own time. In the 1975 revival of Manzotti's ballet, even as dance critics pointed out how much it embodied "the age of European middle class imperialism," or deprecated the saccharine quality of its dancing, people crowded the theater every night, cheering wildly for the triumph of steam power and the victory of reason over ignorance.[21]

NOTES

1. Cyril Beaumont, *Complete Book of Ballets* (London: Putnam's, 1938), is the best single source of information about *Excelsior.*

2. Edwin Evans, "A Romantic Ballet for Christmas?" *Dancing Times* 12 (December 1942): 102.

3. *Il Teatro Illustrato* (Milan), 16 January 1881.

4. "Historic Excelsior Revived," *Dance and Dancer* (April 1975): 38.

5. As quoted in Ivor Guest, *The Divine Virginia [Zucchi]* (New York: Marcel Dekker, 1977), p. 40.

6. Ibid., p. 43. Flaubert's prosecutor, in his brief against *Madame Bovary,* used the same words: "There is in his work no gauze, no veils—it shows nature in the raw." Frederick Brown, *Theater and Revolution: The Culture of the French Stage* (New York: The Viking Press, 1980), p. 32.

7. Mark Edward Perugini, *A Pageant of the Dance and Ballet* (London: Jarrold's, 1946), pp. 229–30.

8. Ivor Guest, *The Empire Ballet* (London: Society for Theatre Research, 1962), p. 31.

9. *Il Teatro Illustrato,* 16 January 1881.

10. Richard Mandell, *Paris 1900* (Toronto: University of Toronto Press, 1967), p. 34.

11. Patrick Geddes, "The Closing Exhibition—Paris, 1900," *The Contemporary Review* 88 (November 1900): 655.

12. See Brown, *Theater and Revolution,* pp. 2–3, for a wonderful evocation of what it was like to be in the Palace of Industry.

13. As quoted in John Allwood, *The Great Exhibitions* (London: Studio Vista, 1977), pp. 61–62.

14. *London Times,* 23 December 1890.

15. *Il Teatro Illustrato,* 16 January 1881.

16. Guest, *Divine Virginia,* p. 36; Beaumont, *Complete Book of Ballets,* p. 638.

17. "Loitering Through the Paris Exposition," *Atlantic Monthly* 65 (March 1890): 372. For an extended treatment of the theme of imperialism and racism at the fairs, see Robert W. Rydell, *All the World's a Fair: Visions of Empire at American International Expositions, 1876–1916* (Chicago: University of Chicago Press, 1986).

18. *Il Teatro Illustrato,* 16 January 1881.

19. "Loitering Through the Paris Exposition," p. 369.

20. His actual words were, "bete au-dela de toute expression." Freda Pitt, "Ballet in Italy," *Dancing Times* 45 (April 1975): 364.

21. "Historic Excelsior Revived," *Dance and Dancer* (April 1975): 38.

The First Generation: Usher, Mumford, and Giedion

ARTHUR P. MOLELLA

The history of invention as an area of formal study dates back at least to the fifteenth century. Early studies concentrated on chronicling developments in invention and engineering and presenting practical methods and taxonomies for the use of working engineers. A vigorous scholarly tradition had its inception in the late nineteenth and early twentieth centuries when such German scholars as Ludwig Beck, Franz Maria Feldhaus, and Conrad Matchoss developed sources and methods for focused internal studies in the field. Then in the 1920s a new approach began to emerge. This essay explores how three historians journeyed beyond internal issues of engineering and invention to consider large social questions regarding the evolution of technology in Western civilization.

In the spring of 1929, Harvard historian Abbott Payson Usher, having turned from conventional economic history to the history of technology, published *A History of Mechanical Inventions*. At precisely the same time, Lewis Mumford finished the final draft of *The Brown Decades* and began to think about a subject for his next book. Within months he was engaged in elaborating the role of the machine in civilization; five years later he produced his own masterpiece, *Technics and Civilization*. Simultaneously, a Swiss art historian named Sigfried Giedion began work on a book tentatively entitled "Konstruktion und Chaos," envisioned as a history of technology in Western civilization, which he subsequently published as *Mechanization Takes Command*. Thus, at the end of the 1920s, at the very same time that the shortcomings of technological society were first becoming evident, these three were first subjecting that society to serious historical scrutiny. Their landmark studies would profoundly influence the development of the field of the history of technology.

From the distance of two generations, these three seemingly spontaneous products still stand out as singular creations. Yet, even though these books were without precedent, they were not isolated phenomena, for they were conceived in a period of profound social and intellectual ferment. The two Americans, having been nurtured in an age of Progressive reform, came to maturity at a time of challenge to traditional dualistic assumptions about knowledge, society, and morals. Dualism implied a strict separation between the soul and nature, mind and body, theory and practice, humanism and science. Such values as Puritan self-denial, traditional education, and elite culture were being reexamined by John Dewey and other pragmatic thinkers.[1] Although Mumford emerged as a stern critic of Deweyite pragmatism, he adhered to an experiential approach similar to Dewey's. A related philosophical transformation underway in Europe was reflected in such movements as modernism and in institutions like the Bauhaus.

Mumford, Usher, and Giedion reflected the values of their times. They were keenly interested in the world of experience and in reuniting disparate aspects of that world rather than breaking it down analytically into specialized bits. Unsympathetic to elitist culture, they argued for an assertion of pragmatic, democratic values. The modern world offered opportunities of self-realization, the reintegration of the individual, concrete holistic understanding, spiritual regeneration, and a new morality. Influenced by their milieu, Usher, Mumford, and Giedion assumed an experiential, experimental, even sensual stance toward the world, disposing them to find pleasure in concrete and tangible aspects of their environment. This stance is conspicuous in the passionate Mumford and Giedion, somewhat less so in the restrained and scholarly Usher.

The history of technology had traditionally dealt with an inert world and artificial objects. Usher, Mumford, and Giedion infused life into this tradition by exploring the interdependency between living beings and dead objects. The revolutionary writings they published or conceived in 1929 showed concern for the whole of matter and spirit. This concern can be demonstrated by examining their works on three levels of relationship—matter to the individual, technology to society, and technology to nature and the cosmos.

A MATERIAL SENSE OF THINGS

The fundamental issue for the history of technology concerns the struggle for material existence. Karl Marx defined this struggle with his revolutionary theories and materialist interpretation of history, exerting a strong influence on late nineteenth-century European economic and social history. His controversial ideas generated an appreciation of physical and material causes but deemphasized traditional political and diplomatic history. Although Marxist history exerted less influence on the more politically conservative Americans, it provided an important backdrop for the writings of Usher and Mumford, if only as something to react against.[2]

Abbott Payson Usher (1883–1965), the son of a prominent Massachusetts lawyer and industrialist, studied economic history at Harvard, later teaching at Cornell University and Boston University.[3] In 1922, he became an assistant professor in the Harvard economics department, where he remained until 1949. Through such European-trained colleagues as Edwin Gay and F. W. Taussig, Usher explored the methods of French and German economic and social historians with their consideration of climatic and geographic effects on social relations. But he objected to their deterministic conclusions. Usher did not view humanity as a passive victim of environmental forces or any other forces—a compelling reason for his denunciation of Marxist philosophy.[4]

Economic geography and other deterministic concepts, Usher maintained, neglected or discounted the effects of human agency. Technology was the "most important fact in the active transformation of environment by human activity."[5] Taking aim on environmental determinism, Usher devoted ten years of intensive study to the history of technology; the result was A History of Mechanical Inventions.

Usher's convictions had deep roots in his democratic political philosophy. Describing himself as an economic liberal, he accredited human agency and specifically the power of the ordinary individual to change his or her environment through technological innovation.[6] But he maintained that his assumptions were scientifically grounded. A pioneer of the "new economic history" and advocate of the "empirical" method, Usher defined a scientific approach to technical history that was midway between the "transcendentalists," who ascribed an almost mystical role to great inventors, and the mechanicomaterialists, who denied the efficacy of human initiative.[7]

Usher opposed ideology, dogma, and untested historical assumptions, developing a reputation for stubbornness among his Harvard colleagues.[8] Preferring the surer knowledge of empirical discovery, he emphasized the role of tangible factors, such as the humidity in English textile mills during the Industrial Revolution.[9] Such tangibles had to be considered, not in isolation, but in relation to others. He examined societies empirically as whole organisms on the assumption that the whole is more important than any of its parts. Usher appreciated the historical role of material factors, especially the concrete qualities of technology, but insisted on fully integrating dynamic human initiative into an otherwise static material world.

Lewis Mumford (b. 1895) estimated material factors differently than Usher. In one sense, he believed in environmental determinism. Further to the left politically than Usher and considerably more respectful of Marx, he placed less emphasis on the individual in the environmental struggle and more on society and the environment itself.[10]

Mumford thought of materials in a very direct, experiential way, a reflection of his activist background in regional planning and architectural history. Extending the notions of his Scots mentor Patrick Geddes,[11] Mumford introduced his three now-famous materially based industrial phases: the eotechnic, associated with wood; the paleotechnic, with iron; and the neotechnic, with lighter metals, alloys, and synthetics. The determining power of material settings was also presented through powerful metaphors. For instance, the "valley section" illustrated how civilizations naturally adapt to material circumstances.[12] The profile of a typical European or North American mountain and river system, the valley section provided a graphic summary of the relationship between occupations and geographic features. At the mountaintop one finds workers in quarries and mines; at the forest level, hunters making tools and weapons.

An even more dramatic and persistent metaphor of *Technics and Civilization* was the mineral mine, the shaping force of the paleotechnic mind. Mumford had been personally impressed by the "dramatic reconstructions of mines" he saw at the Deutsches Museum and in Chicago's Rosenwald Museum (now the Museum of Science and Industry).[13] The dark inorganic environment mysteriously induced the bleak antihuman spirit of the Industrial Revolution. Mumford also suggested a direct comparison between the world of the miner and the reductionist picture of nature painted by Galileo, Newton, and other mechanical philosophers of the seventeenth century.[14] Moving from the dark to the light, the neotechnic world of transparent materials evoked a more humane, enlightened spirit.

Mumford's approach to life and the world showed the influence of Geddes, who encouraged his efforts toward self-realization and moral and material discovery.[15] Frank discussions of sexuality, for instance, pursued the integration of the emotional, physical, and intellectual aspects of human nature.[16] Mumford attempted to act out these ideals in his work and personal life.

Mumford's orientation was toward the tangible, the concrete; he trusted the evidence of the senses in search of aesthetic appreciation. "My sense of the esthetic pleasure of fine machine design," he recalled, "came when I turned a metal standard for my crystal set with none of the conventional flutings and bulgings that were then common."[17] He valued his training at New York's scientific and technical Stuyvesant High School in woodworking, turning, forging, and casting.[18] "Books cannot take the place of first-hand observation," Mumford wrote. Direct experience, especially as a worker, revealed more of the world than "thinking with abstractions and fragments which are as difficult to unify by the methods of specialism as were the broken pieces of Humpty-Dumpty."[19]

Even in the absence of opportunities for "open-air observation," Mumford still denounced abstract preconceptions, recommending visits to industrial museums as a secondary means of knowledge. Museums fascinated Mumford as places combining intellectual with visual and artistic experiences. While researching *Technics and Civilization*, he toured national technical museums in Paris, London, Vienna, and Munich, where he pored over the artifact collections, especially those of the Deutsches Museum.[20] Photographs of these artifacts played a strong documentary role in supporting Mumford's historical theories. In his approach to the material world, he was clearly a visual as well as a verbal thinker.

The third historian under discussion, Sigfried Giedion (1888–1968), emerged from a European background in art and architecture.[21] Born to Swiss-Jewish parents in Prague, site of a family textile mill, Giedion first trained to be an engineer at the University of Vienna. He dutifully finished his engineering studies but felt himself drawn to artistic fields, receiving a Ph.D. at Munich under the famous Swiss art historian Heinrich Wölfflin. Giedion's doctoral thesis and early books addressed the evolution of architectural styles.

Giedion's interests in technology developed from associations with modern artists, sculptors, and architects, including members of the Bauhaus and the C.I.A.M. (Congrès Internationaux d'Architecture Moderne).[22] His *Space, Time, and Architecture,* published in 1941, established him as an influential propagandist for the Bauhaus and for modernism in art and architecture. The Bauhaus school's call for the unification of art and technology as well as handicraft and mass production echoed throughout "Konstruction und Chaos," his first approach to technological history, and even more insistently in *Mechanization Takes Command.*[23]

Mumford and Usher dealt extensively with the character of the raw physical environment, but Giedion primarily addressed the constructed environment of objects, handicrafted as well as mass produced. He believed that objects directly influenced the "thoughts" and "feelings" of a culture, affecting people emotionally and artistically.[24] Giedion's goal was to discover the essence, the "constituent" elements, of material things—the "spirit of the age." The configuration of a chair, the accouterments of the bath, even the structure of a lock shaped the human consciousness. *Mechanization Takes Command* was emblematic of the artistic powers of manufactured goods. Juxtaposing images of paintings and sculptures with telling engravings of slaughterhouse "disassembly lines," the Bauhaus-designed book strikes one as a work of modern art in its own right.[25]

Giedion, like Mumford, insisted on the importance of first-hand observation, the direct experience of materials, and general reliance on evidence of the senses, all in keeping with the Bauhaus school's pedagogical theory.[26] Giedion visited Britain's museums in South Kensington, conducted personal visits to slaughterhouses, flour mills, and industrial settings, and mounted his own exhibits.[27] In 1935, at the Zürich Museum of Arts and Crafts, he curated an exhibition, "The Bath Today and Yesterday," one of the unique subjects later considered in *Mechanization Takes Command.*[28]

Giedion's strategy in the history of technology was to emphasize banal particulars normally overlooked by conventional histories. In discovering a spiritual essence in these objects, he assumed a direct communion between humans and the material world, reaffirming the modern spirit of inquiry.

THE HISTORY OF TECHNOLOGY AS SOCIAL HISTORY

Usher, Mumford, and Giedion all perceived a new relationship between humanity and the material environment. This awareness of human interaction inevitably evolved into a concern with social history. Having established a methodology of material interaction, they aimed to understand social history from a material perspective.

Lewis Mumford tackled the critical cultural problem of the day, the social and spiritual ills of the Machine Age. At issue was the apparent conflict between spiritual and material values during the era of automation, mass production, "Fordismus," and Coolidge prosperity and mass consumption.[29] Some welcomed these developments as signs of material and social progress, while pessimistic observers voiced their fears of the brutalizing and dulling effects of mechanization and pointed to the deadly applications of technology in the World War.

Mumford joined the ongoing machine debates with such intellectuals as Van Wyck Brooks, Joseph Wood Krutch, Charles Beard, and Stuart Chase.[30] He expressed mixed feelings toward the cultural effects of technology. Although deeply critical of the character of past and current technological societies, Mumford was fascinated with and hopeful about technology and the future. The opinions of Thorstein Veblen, that trenchant critic of industrial capitalism, and again of Geddes influenced his thinking.[31] At the New School for Social Research and as an editor of the *Dial,* Mumford nurtured his own left-wing communitarian views of the industrial process.

Mumford's purpose in writing *Technics and Civilization* was to place current issues in historical context, but the book was as much polemics as history. Describing it to Van Wyck Brooks as a "grand broadside at the mechanists and the mammonists from within their own citadel," he claimed his new book was one to "out-Bentham the Benthamites, to out-Marx the Marxians, and in general, to put almost anybody and everybody who has written about the machine or modern industrialism or the promise of the future into his or her place."[32]

Mumford was interested in the machine as social process and accordingly dealt with "technics," the socially diffused practice of the industrial arts, rather than with "technology," the rational study of those arts.[33] He described the machine as the product of multiple social choices over centuries. Yet Mumford was not the social historian Usher was, and he preferred abstractions to studies of individual people. The machine became "The Machine," taking on a life of its own in "The Drama of the Machines," his first small publication on the history of technology.[34] He considered broad phenomena, cultural "moods," social psychology, and the "spiritual contributions" the machine has made to culture. "Form and Personality," the title originally given to *Technics and Civilization,* indicated this abstract socio-psychological perspective.[35]

Whether Mumford thought society influenced technics or vice versa is debated. Actually, the influence appeared to be mutual. For example, following the German historian Werner Sombart, Mumford argued that capitalism with its demands for social regimentation begot technology.[36] In turn, inventions such as the clock and the steam engine reinforced the economic and social order. The effect was mutual reinforcement.

Giedion was also concerned with technology as a fundamental sociocultural phenomenon. *Mechanization Takes Command* was only one part of an envisioned multivolume work, "Die Entstehung des heutigen Menschen" (The Origins of Modern Man), dealing synoptically with the social history of Western culture.[37] The unpublished "Konstruction und Chaos" was to be the first installment, but the project was interrupted by Giedion's trip to America in 1938.[38] Instead, *Space, Time, and Architecture* appeared and eventually its sequel, *Mechanization Takes Command.*

Giedion was uninterested in great events or the lives of politicians, diplomats and military heroes, preferring what he termed "anonymous history," the lives of the humble makers of humble things.[39] He placed special emphasis on the pervasive artifacts of mass production, including chairs, kitchen appliances, bathroom fixtures, and other domestic accouterments, because he believed that commonplace objects harbored important cultural information—a conviction expressed in his aphorism, "the sun is mirrored even in a coffee spoon."[40] Giedion derived this conception of history not only from Wölfflin but from artist friends whose paintings "taught [him] to observe seriously objects which seemed unworthy of interest" and "have shown in their pictures that . . . the unnoticed articles that result from mass production . . . have become parts of our natures."[41]

Giedion urged historians to leave their cloistered world and participate in efforts to improve society.[42] In *Mechanization Takes Command* he hoped to use history to expose and heal the cultural wounds inflicted by mechanization. If mechanization had caused a schism by dominating culture, it was time for human beings to reassume cultural control through the merger of thought and feeling, of industry and art, of reason and emotion. Although these ideas had been propounded by the Bauhaus a decade earlier, Giedion brought them to the attention of the American audience.

Giedion's visit to the United States in 1938 marked a turning point in his career and in his approach to technological history. Finding his heterodox views unwelcome among Swiss art historians and in Nazi-dominated Germany, Giedion eagerly accepted Walter Groupius's invitation to deliver Harvard's prestigious Norton Lectures, which became the basis for *Space, Time, and Architecture.*[43] Although he admired American inventiveness, Giedion was astonished by the degree to which mechanization dominated American culture and by its glaring negative effects. Giedion transformed the original

European-based "Konstruktion und Chaos" to reflect his American findings.[44] He pursued such previously untouched sources as patent specifications and records of obscure companies, regarding the United States as a vast laboratory for comparative tests of the effects of mechanization on culture.

Of the three historians, Abbott Payson Usher developed the least abstract, most serious and suggestive ideas about technology and social history. Although historians today tend to regard him as a sort of internalist par excellence—the result of contemplating the exquisite detail of his analyses of mechanical processes—it is clear that what is most important about *A History of Mechanical Inventions* is its presentation of a broad strategy for social history. Usher after all was a social historian at heart. The study of technology, he thought, yielded important clues to understanding social change.

Usher developed his argument for technical history in almost syllogistic fashion. History was complex and ambiguous, but Usher was certain that societies undergo massive structural changes over time. He wanted to know the reasons for these structural transformations. History is essentially dynamic, and the cause of this dynamism is innovation, the generation of novelty. Technology, he believed, epitomized the innovative process. Technological sequences, without necessarily implying progress, only make sense "when read forward." Technology, therefore, is "deeply imbedded in time" and "among the most intense manifestations of the dynamic processes of history."[45] Moreover, technological processes are relatively easy to trace and can be examined under concrete and objective circumstances, at least when compared with innovations in humanistic areas such as religion or the arts. He contrasted his dynamic views with those of economic geographers and others who stressed static environmental influences.

One of Usher's tools for explaining technological innovation as social process was the German theory of gestalt psychology.[46] Usher recognized the power of this concept from his Cornell colleague, the psychologist Robert Ogden.[47] Denying that gestalt was a metaphysical concept, Usher considered it an empirical description of human behavior, far less mysterious than alternative psychological theories.[48] His well-known four-stage conception of gestalt served his historical methods by demonstrating that the gestalt problem-solving ability was innate in everyone, and therefore any member of society represented a potential source of social change. The ability to complete patterns was not a rational but rather a common and primitive visual/emotional process. Inventive geniuses create in the same way as everyone else, simply showing the ability to a higher degree. They are not, in fact, a source of historical discontinuity. This view reinforced Usher's belief that history is a continuous process, with occasional nodal points of rapid change.[49]

A History of Mechanical Inventions is theoretically suggestive, but, overall, it contains very little in the way of social history, appearing rather as the epitome of internalism. The ideas on social history and gestalt are restricted to the first two chapters, which became four in the book's second edition in 1954. The narrative chapters are almost devoid of social, economic, or psychological considerations. What gives the book this split personality?

The answer lies in Usher's analytical method of multilinear analysis outlined in the second edition of his book. Using graphic illustrations, Usher demonstrated that history actually consists of separate lines of development called "systems of events," independent sequences of causally linked historical events.[50] Technology constituted one such general system, consisting in turn of multiple subsystems. Because of the complexity of historical causes, it is important to conduct separate analyses of each system over a long time. Justifying the application of this methodology in *A History of Mechanical Inventions,* Usher explained:

> There is special value, too, in a separate study of the technological problems without regard for the moment to the detailed economic consequences of particular changes. The development of all these ultimate consequences would require a whole series of separate and measurably independent narratives.[51]

The final stage of analysis was to compare the separate systems in cross-sections of time. The result of this deliberate, perhaps overly cautious approach was only suggestive strategy; it did not, in itself, provide a model for the social history of technology. (Ironically, the less cautious Giedion and Mumford managed to produce evocative social histories, despite their lack of a rigorous method.) Still, Usher's method led him to write one of the first serious studies of technology, which most conventional economic historians had either treated as an exogenous variable or caricatured with uncritical accounts of heroic inventors. His method was also evocative in spirit, if not in practice, of the modern discipline of social history. *A History of Mechanical Inventions* caught the attention of Marc Bloch and his fellow *Annalistes,* who conducted interdisciplinary studies of social phenomena over long time periods and were equally fascinated by the cultural characteristics of technology. Usher had developed his method before and apparently independently of the rise of the *Annales* school, although he doubtless drew on many of the same sources and historical models.[52]

In venturing into social history, Usher, Mumford, and Giedion viewed the history of technology not as a new specialty, contributing further to the fragmentation of knowledge, but as a synthesis of ideas from specialities as diverse as psychology, sociology, anthropology, art, science, and economics. This thinking ran counter to the tendency toward increasing specialization, one reason for Mumford's disdain of conventional academics. Giedion was a vehement critic of specialization, believing it fragmented our understanding of humanity.[53] He sought the cooperation of like-minded Americans at Harvard, Massachusetts Institute of Technology, and the University of Chicago in establishing chairs of interdisciplinary studies under an Institute for Contemporary History and Research, dedicated to "anonymous history." The proposal was apparently never realized.[54]

MECHANIZATION AND THE PHILOSOPHY OF NATURE

The three philosophically minded historians related the unity of knowledge to the natural order, currently shifting away from prevailing mechanical modes. They criticized the seventeenth-century mechanical philosophy and its extensions in the modern world, considering it a fragmented view that reduced nature to a machine. Objectifying the world in terms of matter and motion, it tore the human mind and soul out of nature, giving rise to an insidious dualism between matter and spirit afflicting modern culture. The mechanical philosophy engendered the evils of technology and mechanization.

Mumford, Giedion, and Usher followed developments in modern physics and biology, developments that seemed to portend the overthrow of the old mechanical world view. None of them had significant contacts with the scientific community, but they did avidly read the popular scientific literature as well as the philosophy of science. The science popularizer Edward Slosson introduced Mumford to the writings of Ostwald, Pearson, and Mach,[55] whereas Giedion gained many of his insights from painters, sculptors, and architects.[56] All three historians discerned a trend toward unity, interdependency, and structure in contemporary science and philosophy. They traced this change to modern physics, especially relativity and quantum mechanics. Such theories placed the human mind back in nature by demonstrating that, in the act of observation, the observer disturbs the natural process. Mind and matter were one at some level of reality. Physics replaced matter and its properties with structures and patterns of events in space-time, allowing Giedion and Mumford to unify matter and spirit scientifically.

In *Space, Time, and Architecture,* Giedion played with ideas of space-time, the fourth dimension, simultaneity, and new concepts of matter, relating these physical ideas to what his artist friends achieved in painting or sculpture.[57] His research notes include such relativistic speculations as "a moving body carries with it a temporal system, entirely different from that of the world," and "the true units of time and space are neither points nor moments but moments-in-the-history-of-a-point." Giedion then attempted to reconcile such concepts with Kantian categories of space and time.

In his unpublished writings, Giedion speculated that the healing of the culture would begin within the special sciences, especially modern physics and biology, where "we find a departure from the investigation of an isolated process, from purely mechanistic conceptions of the world." Thanks to modern physics, the "cosmos is beginning to resemble more and more one great thought than a big machine."[58] Perceived as clearly by cubist painters as by abstract scientists, this vision represented the fulfillment of the modernist dream of uniting art, technology, and science.

Mumford disagreed with those who believed that science had destroyed metaphysics and religion, optimistically observing that "the painfully limited science and metaphysics of the last three hundred years are undergoing a profound transformation, which will modify all the current conventions of thought."[59] Usher contended that modern science offered a new, more profound way of understanding history, for empiricist methods could now grasp the complex interdependencies of human social behavior.[60]

Although originating in physics, the new concepts of interactivity invited biological forms of explanation. The three historians of technology viewed history as an essentially biological process and technology as a form of biological adaptation.[61] Giedion detected a growing tendency away from mechanistic conceptions toward organicism in scientific thought. Basing the neotechnic age on such organic models, Mumford noted in *Technics and Civilization* that "physiology became for the nineteenth century what mechanics had been for the seventeenth: instead of mechanism forming a pattern for life, living organisms began to form a pattern for mechanism."[62]

These historical insights reflected major trends in contemporary philosophy and biology. In *Science and the Modern World* (1925), Alfred North Whitehead advanced an influential theory of "organic mechanism."[63] Whitehead perceived a radical transformation in the foundations of physics, in which holistic organic models supplanted mechanical principles. Whitehead based this holistic view on the idea that the whole is more than the sum of its parts. Thus, the electron within the body is different from the electron outside the body by virtue of the body's organization. Although the term *holism* had been coined by Jan Smuts, the Prime Minister of South Africa, only in 1926 (in *Holism and Evolution*), the concept was already widespread.

Whitehead had a direct influence on the emergence of the holistic organicist movement in twentieth-century biology, itself a reaction against mechanistic reductionist traditions. Inspired by Whitehead, Harvard's L. J. Henderson developed holistic conceptions of physical chemistry and physiology, participating with biologists such as J. S. Haldane and J. H. Woodger.[64] An influential neovitalist version of holism was also expounded by the French philosopher Henri Bergson.[65]

Holism captured Mumford's interest before he wrote *Technics and Civilization*. In correspondence with a friend, he reported his impression of the novel organicist ideas of Henderson.

> My latest discovery is L. J. Henderson, whose book on The Fitness of the Environment is really a revolutionary contribution to biology, and

for that matter to philosophic thought, for, by ingenious improvements on the old arguments of Paley and the natural theologians of his time, he has restored the concept of design in Nature, and I think definitely proved that the physical environment is a biocentric one![66]

A direct influence on the holistic ideas in *Technics and Civilization* was Patrick Geddes and J. Arthur Thomson's *Life: Outlines of General Biology,* a neovitalist tract replete with holistic notions of physics, biology, and society. Incorporating the ideas of Smuts, Einstein, and Whitehead, the authors described the universe as a "vibrant web of interrelations," a synoptic, nonmechanical whole, unified by the effects of evolution and the "general idea of relativity."[67]

Mumford's holistic view, of course, did not imply rejection of the machine. Fascinated by technology since childhood and at one time having even contemplated a career in electrical engineering, he had no desire to take the technics out of civilization. Rather, holism meant to him the total assimilation of the machine into society on a new, organic basis. Mumford elaborated his central thesis of *Technics and Civilization* in further correspondence.

Up to the neotechnic period technological progress consisted in renouncing the organic and substituting the mechanical. This reached its height around 1870. Since then the new trend, visible in technics as well as in philosophy as in social life, is the return to the organic by means of the mechanical: a return with a difference, namely with the whole body of machines and analytical knowledge we have acquired along the way.[68]

Gestalt principles, holistic by definition, harmonized with the new ways of conceptualizing nature. Developed in reaction against mechanistic associationist psychology, gestalt involved the entire organism, body as well as mind, in the interpretation of nature.[69] Usher defended gestalt as an empirical method but invested it with higher meaning. Not only was it a means of unifying physical and mental processes, it also provided a foundation for a revised conception of society.[70] One again sees the direct linkages between Usher's ideas of science and technology and his political philosophy.

Usher's recognition that individuals were interdependent led him to modify the concept of economic liberalism from strict adherence to laissez-faire principles to acceptance of more state intervention in society.[71] Laissez-faire, Usher reminded us, was grounded in eighteenth-century ideas of natural law, mechanistic conceptions of "an orderly and rational system of nature." In contrast, holism and the new natural philosophy replaced "mechanical categories of cause and effect" with "concepts of adaptation, and of reciprocal determination." Society, like nature, was less competitive and more cooperative than previously believed.

Giedion thought that gestalt and holism were two sides of the same coin, both antidotes to the disequilibrium between thought and feeling that threatened to destroy modern machine civilization. Research notes for *Mechanization Takes Command* dwell on the holistic implications of gestalt in such passages as, "biology today gives an urgent warning in the Gestalttheorie, the theory of compounds, that 'the whole is more than the parts.' "[72] Positing a structural continuum between mind and matter, Giedion wrote that "Holism regards the chemical patterns incorporated into the biolog[ical] patt[erns] and both of these into the subsequent psychical patterns or wholes." From this he deduced the universal structural principle that "the central feature or character of the cosmic movement is . . . towards wholeness or holeness."[73]

An ideal of holistic history was the ultimate implication of such viewpoints. Mumford and Giedion, especially, perceived relationships between emergent cultural phenomena, suggesting an underlying unity. In *Technics and Civilization,* for example, we discover that the motion picture coordinates time and space on their own axis and introduces the observer into the image through the shifting eye of the camera. This popular art form expressed the modern world-picture and ideas of space and time, which "are already part of the unformulated experience of millions of people, to whom Einstein or Bohr or Bergson . . . are scarcely even names."[74] Likewise, Giedion remarked that the concept of simultaneity was at issue in both contemporary art and theoretical physics and that the physics of Newton shared some baroque qualities with the art and architecture of its period. Such similarities manifested the "unity of science, art, and life"—that is, the "unity of culture."[75] Giedion's determination to trace these kinships to the Spirit of the Age reveals his ultimate debt to the German Idealist tradition, transmitted through Wölfflin and the Swiss Renaissance scholar Jacob Burckhardt.[76] But the historical value of *Mechanization Takes Command* lies in its attempts to perceive the universal in the particular—in the coffee spoons, chairs, and other objects of everyday life.

Science and the holistic view dematerialized matter, replacing its concrete properties with structures and transcending the dualistic boundary between matter and spirit. This new philosophy of nature permitted Usher, Mumford, and Giedion to develop a history of technology incorporating notions of unity and interdependency.

CONCLUSION

In writing their histories of technology, Usher, Mumford, and Giedion faced the central problem of their generation: the mechanization of the Western world. Their large-frame histories encompassed the moral and human dimensions of this problem. Mumford, the cultural critic, sought to reform the structure and personality of society. Giedion, influenced by European modernism, aimed to reform art and technology. Reform was always a component of economics, and Usher sought to modify traditional notions of laissez-faire liberalism by appealing to the principles of a nonmechanical world order. All three recognized that such questions were deeply rooted in time. In their view, the eternal conflict between matter and spirit represented the crucial issue for the history of technology. Humanists above all, they approached the history of technology in a way that did not diminish human beings. This approach meant treating technology in social context.

Of the three, Usher presented the most positive vision of technology: not an oppressive, antihuman force but a means of liberation from brute material conditions. Giedion, at once fascinated and threatened by the machine taking command, pleaded for a regeneration of feeling and humanity in the face of uncontrolled technology. Mumford portrayed mankind as basically at the mercy of machines and mechanized civilization, but found hope in a new organic, holistic concept. All three believed that they were about to witness a major cultural transition from a mechanical to an organic world picture. In their histories of technology, they attempted to measure the implications of this transition for understanding the present as well as the past.

The appearance of these three classic histories was in itself a historical phenomenon: they were the result of particular conditions, a mood, a particular time and place. Their approach to the history of technology reflected this ambience, embracing concrete experience, tactile understanding, and self-fulfillment through direct exploration of the environment.

This optimistic and experimental mood challenging traditional authority was shattered by World War II. The history of technology became a formal academic specialty in the late 1950s, a more conservative era. The field went through a period of intense internalism, much as had the history of science.[77] In the late 1960s, many historians felt dissatisfied, perhaps embarrassed, with this narrow viewpoint and revived a contextual approach. They have produced an impressive literature, and yet it clearly differs from the work of Usher, Mumford, and Giedion in that the focus is on the context of very narrowly defined problems. Truly, every generation writes its own history. The books of Usher, Mumford, and Giedion have remained classics. That they endure is due less to the answers they gave to historical problems than to the eternal quality of the questions they pose.

NOTES

For help with this essay, I wish to thank Robert C. Post, Miriam Usher Chrisman, Roger Sherman, Philip J. Pauly, Ruth Maus, Elisabeth Larson, and Dorothee Huber.

1. See John Dewey, *Experience and Nature,* in *The Later Works,* vol. 1, ed. Jo Anna Boydston (Carbondale, Ill.: Southern Illinois University Press, 1981); *Art as Experience* (New York: Capricorn Books, 1934); "Affective Thought," in *Philosophy and Civilization* (New York: G. P. Putnam's, 1931), pp. 117–25. For Mumford's

criticisms of Dewey, see Gary Bullert, *The Politics of John Dewey* (Buffalo, N.Y.: Prometheus, 1983), pp. 82–85, 176–82.

2. *International Encyclopedia of the Social Sciences,* s.v. "Marxist Sociology." Gustav Schmoller and Werner Sombart both transmitted and transformed Marx's historical ideas. For the American response, see William N. Parker, "Historiography of American Economic History," *Encyclopedia of American Economic History* (New York: Scribners, 1980). See the reactions of Usher's teacher and colleague Edwin F. Gay to Marxist determinism in Herbert Heaton, *A Scholar in Action: Edwin Gay* (Cambridge, Mass.: Harvard University Press, 1952), p. 38.

3. *Harvard University Gazette,* 12 March 1966, pp. 155–56; "Memorial," *Technology and Culture* 6 (Fall 1965): 630–32; *National Cyclopaedia of American Biography* (New York: James T. White & Co., 1930). I am grateful to Miriam Usher Chrisman for biographical information on her father.

4. Abbott Payson Usher, *A History of Mechanical Inventions* (New York: McGraw-Hill, 1929), p. 2.

5. Ibid., p. 1.

6. "The 'History of Mechanical Inventions' was designed to illustrate the action of the individual in innovation, not only in technology and science, but in many other fields" (student lecture notes taken by Hugh Aitken, 21 October 1948; personal files of Dr. Aitken).

7. Abbott Payson Usher, "The Significance of Modern Empiricism for History and Economics," *The Journal of Economic History* 9 (1949): 137–55; *A History of Mechanical Inventions,* rev. ed. (Cambridge, Mass.: Harvard University Press, 1954), pp. 18, 61.

8. *Harvard University Gazette,* 12 March 1966; Miriam Chrisman recalls that Usher's insistence on standing by a position he believed in made him unpopular with the Harvard administration.

9. Abbott Payson Usher, *An Introduction to the Industrial History of England* (Boston: Houghton Mifflin: 1920), pp. 263–65.

10. Lewis Mumford, *Technics and Civilization* (New York: Harcourt, Brace and Co., 1934), pp. 110, 464.

11. Ibid., pp. 109, 110. Geddes attempted to recruit Mumford as a collaborator. (Geddes to Mumford, 25 February 1921; Mumford Papers, University of Pennsylvania Library, Philadelphia).

12. The "valley section" image was borrowed directly from Geddes. *Technics and Civilization,* pp. 60–64; Patrick Geddes and Arthur J. Thomson, *Life: Outlines of General Biology,* vol. 2 (New York and London: Harper, 1931), pp. 1395–98; Paddy Kitchen, *A Most Unsettling Person: The Life and Ideas of Patrick Geddes* (New York: Saturday Review Press, 1975), p. 137.

13. *Technics and Civilization,* p. 447.

14. Ibid., pp. 20, 69, 70.

15. One can also discern the influence of Dewey, despite Mumford's public disagreement with him. Geddes's notions of learning by experience can be seen in Geddes and Thomson, *Life,* 2:1402.

16. Throughout the Geddes-Mumford correspondence, e.g., Mumford to Geddes, 8 September 1920, Mumford Papers. Also see Mumford's autobiography, *Sketches from Life* (New York: Dial Press, 1982), p. 155.

17. Mumford to Melvin Kranzberg, 20 December 1969, in Kranzberg's personal files. Much of the material in this letter is published in *Technology and Culture* 11 (April 1970): 205–13, on the occasion of Mumford's receipt of the da Vinci Medal.

18. Ibid.

19. *Technics and Civilization,* p. 447.

20. Ibid., and Mumford to Kranzberg, 20 December 1969.

21. Autobiographical manuscript, Giedion Papers, Institut für Geschichte und Theorie der Architektur, Eidgenössische Technische Hochschule (ETH) in Zürich, Switzerland; Obituary, *Werk* 55 (1968): 337–38; Giovanni Koenig, "Ricordo di Siegfried [sic] Giedion," *Casabella* 32 (April 1968): 89–90; Obituary, *New York Times* 12 April 1968, p. 35; Obituary, *Design* 234 (June 1968): 83.

22. Autobiography, Giedion Papers, ETH. Giedion was a founder and long-time General Secretary of the peripatetic C.I.A.M.

23. Sigfried Giedion, *Space, Time, and Architecture: The Growth of a New Tradition* (Cambridge, Mass.: Harvard University Press, 1941). The manuscript of "Konstruktion und Chaos" survives in the Giedion Papers, ETH. I am deeply indebted to the then-archivist of the Giedion Papers, Dorothee Huber, for general assistance with his papers, particularly for background information on the uncompleted book. See her " 'Konstruktion und Chaos': il grande progetto incompiuto," *Rassegna* 25 (1986): 62–72. Sigfried Giedion, *Mechanization Takes Command: A Contribution to Anonymous History* (New York: Oxford University Press, 1948). For the ideology of the Bauhaus, see Hans Wingler, *The Bauhaus: Weimar, Dessau, Berlin, Chicago* (Cambridge, Mass.: MIT Press, 1969).

24. *Mechanization Takes Command,* pp. 2–11.

25. The book was designed by Herbert Bayer, Bauhaus typographer. Rainer Wick, *Bauhaus Pädagogik* (Köln: DuMont, 1982), p. 43.

26. Giedion stated that he was always concerned with "direct contact with [the] eye" (Giedion autobiographical manuscript, in the Giedion Papers, ETH). The linking of craft with pedagogy was part of a long tradition running from Locke through several Swiss educational reformers, including Pestalozzi, Rousseau, and Fröbel. See Wick, *Bauhaus Pädagogik,* p. 64.

27. Giedion's correspondence in 1942 and 1943 documents his personal inspection visits; e.g., Lucy Porter to Giedion, 27 February 1942, concerns his research on the origins of the Yale lock. Giedion Papers, ETH.

28. Documented in Giedion's correspondence from April 1935 and in his published catalog, available in Giedion Papers, ETH.

29. The literature on mass production and 1920s culture is extensive. See, e.g., David A. Hounshell, *From the American System to Mass Production, 1800–1932* (Baltimore: Johns Hopkins University Press, 1984), pp. 303–30; André Siegfried, *America Comes of Age, A French Analysis* (New York: Harcourt, Brace and Co., 1927); Frederick Lewis Allen, *Only Yesterday* (New York: Harper and Brothers, 1931); Paul A. Carter, *The Twenties in America,* 2d ed. (New York: Crowell, 1968).

30. Richard Striner, "The Machine as Symbol, 1920–1939" (Ph.D. dissertation, University of Maryland, 1982), deals extensively with Mumford and Stuart Chase.

31. Mumford to Kranzberg, 20 December 1969. *Technics and Civilization,* p. 475, acknowledges Geddes, Veblen, and sociologist Victor Branford as the principal influences in writing his book.

32. Mumford to Brooks, 21 June 1933. Mumford Papers.

33. Mumford to Kranzberg, 15 January 1970, in Kranzberg's personal files.

34. *Scribner's Magazine* 88 (1930): 151–61.

35. Draft in Mumford Papers.

36. See Werner Sombart, *Der Moderne Kapitalismus,* 4 vols. (Munich: Duncker and Humblot, 1927).

37. Described in "Konstruktion und Chaos," Giedion Papers, ETH.

38. Giedion's close friend Moholy-Nagy solicited the manuscript for his new Bauhaus series in Chicago. Moholy-Nagy to Giedion, 22 December 1937, Giedion Papers, ETH.

39. In his discussion of "Anonyme Geschichte" in Box "SG-Unternehmungen," folder "Project: Inst. for Contemporary History and Research," Giedion Papers, ETH (hereafter "SG-Unternehmungen").

40. *Mechanization Takes Command*, p. 3. For Giedion's philosophical heritage, see Spiro Kostof, "Architecture, You and Him: The Mark of Sigfried Giedion," *Daedalus* 10 (1976): 192–93.

41. *Space, Time, and Architecture*, p. 4.

42. "SG-Unternehmungen."

43. Personal communication from Andraeus Giedion and Verena Clay, son and daughter of Giedion; correspondence with Gropius, e.g., Gropius to Giedion, 13 April 1938, Giedion Papers, ETH.

44. Compared to the heterodox *Mechanization Takes Command*, "Konstruktion und Chaos" took a conventional approach to the history of industrialization.

45. *History of Mechanical Inventions* (1929), p. 4.

46. "Invention is a social process in which many individuals participate" (Usher, "The Significance of Modern Empiricism," pp. 137–55). Gestalt was just arriving in America in the 1920s (see Michael M. Sokal, "The Gestalt Psychologists in Behaviorist America," *American Historical Review* 89 [1984]: 1240–63). Usher was one the first Americans to apply gestalt principles to a discipline other than psychology.

47. Ogden, Usher's housemate at Cornell, translated and popularized gestalt concepts in America. Gestalt ideas appeared in *History of Mechanical Inventions* (1929), chap. 2, and more fully in the 2d ed. (1954), chap. 4.

48. *History of Mechanical Inventions* (1929), p. 8; (1954), p. vii.

49. Ibid. (1929), p. 6.

50. Ibid., chap. 2. See the analysis of William Parker in his "Abbott Payson Usher," *Architects and Craftsmen in History*, ed. Joseph Lambie (Tübingen: Mohr, 1956), pp. 157–66.

51. *History of Mechanical Inventions* (1929), p. vii.

52. Marc Bloch favorably reviewed Usher's book in *Annales* 2 (1931): 278–79, but I have been unable to locate any correspondence between Usher and the founders of the *Annales d'histoire économique et sociale*. Bloch's most acclaimed article dealt with the cultural diffusion of the water mill, "Avènement et conquêtes du moulin à eau," *Annales* 7 (1935): 538–63.

53. "The outstanding person of our period, and its stigma as well, is the specialist. He grows out of the split personality and the unevenly adjusted man" ("Vorlesungen Yale. Trowbridge Lectures 1941/1942," Giedion Papers, ETH). Giedion's correspondence deals extensively with the evils of specialization. For instance, Moholy-Nagy to Giedion, 4 June 1938; Giedion to Walter Gropius, 21 March 1938.

54. Giedion wrote a prospectus for the projected Institute (in "SG-Unternehmungen") and sent it to John U. Nef at Chicago, President Conant at Harvard, and John Burchard, Dean of Humanities at M.I.T. (Giedion to Burchard, 5 February 1941; Giedion to Conant, 16 November 1940; Giedion to Nef, 3 August 1943. Giedion Papers, ETH.)

55. Also included was an early introduction to relativity theory. (Mumford to Kranzberg, 20 December 1969.)

56. "Notizien, Einleitung und Folgerungen" (to the ms. for *Mechanization Takes Command*), Giedion Papers, ETH. In *Space, Time, and Architecture*, Giedion frequently alluded to parallels between artistic and scientific concepts (e.g., p. 357).

57. Such ideas pervade his reading and research notes to the manuscript for *Space, Time, and Architecture;* see Giedion Papers, ETH, especially box, "Notizen und Manuskripte Teile VI, VII, VIII, IX," Folder "Futurismus."

58. c. 1938, in "SG-Unternehmungen."

59. Mumford to Bernard Smith, 9 May 1929, Mumford Papers.

60. Usher "The Significance of Modern Empiricism," pp. 140, 142–43. Usher had been reading Eddington, Whitehead, and Bergson, among other advocates of the new approach.

61. Ibid., p. 143: "The problems of process open up a number of questions that resemble the analytical problems of the biological sciences." In *Mechanization Takes Command*, p. 508, Giedion noted the "trend toward the organic that asserted itself in the early 'Thirties and gained strength in the following years." The same thought is pervasive in the last two chapters of *Technics and Civilization*.

62. *Technics and Civilization*, p. 216.

63. I have used the New York, 1931, edition of Whitehead, p. 116.

64. John L. Parascondola, "Lawrence J. Henderson and the Concept of Organized Systems" (Ph. D. diss. University of Wisconsin, 1968), pp. 67–129. For the holistic movement in biology, see Garland Allen, *Life Science in the Twentieth Century* (New York: Wiley, 1975), pp. 103–6.

65. See Whitehead, *Science and the Modern World* (1931), p. 212.

66. Mumford to James Henderson, 24 June 1929, Mumford Papers.

67. Geddes and Thomson, *Life*, vol. 1, p. vii; vol. 2, pp. 1114–15, 1376–83.

68. Mumford to James Henderson, 8 August 1933, Mumford Papers.

69. Kurt Koffka, one of the founders of gestalt psychology, makes explicit these philosophical connections in his article, "Gestalt," *Encyclopedia of the Social Sciences* (New York: Macmillan, c. 1930).

70. "Gestalt concepts emphasize the wholeness of the organism." Usher, "The Significance of Modern Empiricism," p. 151.

71. "A Liberal Theory of Constructive Statecraft," Presidential Address Delivered at the Forty-Sixth Annual Meeting of the American Economic Association, 27 December 1933, *The American Economic Review* 24 (1934): 1.

72. "SG-Unternehmungen."

73. In notes to the ms. for *Mechanization Takes Command*. "Notizien, Einleitung und Folgerungen," in notebook "M.T.C. Excerpts, Conclusion," Giedion Papers, ETH.

74. *Technics and Civilization*, p. 342.

75. Notes for "Konstruktion und Chaos," box, "Konzeption und Skizzen," Giedion Papers, ETH.

76. See Kostoff, "Architecture, You and Him," and Marshall McLuhan's enthusiastic review of *Mechanization Takes Command,* in *Hudson Review* 1 (1949): 599–600.

77. See Colleen A. Dunlavy, "Transcending Internalism: On the Historiography of Technology in the United States," unpublished research paper, 1983, and John Staudenmaier, *Technology's Storytellers: Reweaving the Human Fabric* (Cambridge, Mass.: MIT Press, 1985), p. 25.

Machines, Megamachines, and Systems

THOMAS P. HUGHES

CONTRASTING IMAGES AND CONCEPTS

Political, economic, and social experiences, not technological ones, dominate the American memory and images of the past. In the popular mind, the roots of the American character are to be found in New England villages, colonial Philadelphia, Southern plantations, the unsettled frontier expanses of the prairies and the great plains, and the battlefields of Gettysburg. The formative years are believed to be the era of gilded-age business, the "over there" years of World War I, and the roller-coaster decades of the "roaring twenties" and the "dust bowl" thirties. Maturity is thought to have come with the prosperity and military power of the 1950s.

Generations of historians and popular writers have created or sustained these concepts and images. Consider, for instance, the memorable images and phraseology of the highly influential American historian Frederick Jackson Turner who wrote of the influence of the frontier on American history and lastingly shaped our self-image. He asked his reader to stand at the Cumberland Gap early in the nineteenth century and watch the flowing westward of successive waves of Indians, fur traders, hunters, cattle raisers, and pioneer farmers.[1] Turner permanently influenced not only our self-image but our self-definition when he attributed to the frontier experience predominant and long-lived American characteristics. He saw the frontier experience promoting political democracy, the cross-pollination of ethnic groups, and that "coarseness and strength combined with acuteness and inquisitiveness; that practical, inventive turn of mind, quick to find expedients; that masterful grasp of material things, lacking in the artistic but powerful to effect great ends; that restless, nervous energy; that dominant individualism working for

The George Sarton Memorial Lecture sponsored by the History of Science Society at the Annual Meeting of the American Association for the Advancement of Science, May 26, 1986, Philadelphia, Pennsylvania.

good and for evil, and withal that buoyancy and exuberance which comes with freedom."[2]

These are memorable images and popular concepts, but do they capture the essence of the American past that has shaped American character? From the historian of technology, the answer is no. This essay will suggest other memories and images of the American past and another interpretation of the American character shaped by it. Having presented an alternative perspective on American history and character, I shall then suggest fundamental problems and opportunities facing us today that arise from this rereading of history. My mentors in this engagement with the profound and historical influences of technology and society will be Lewis Mumford, Sigfried Giedion, and Jacques Ellul, all of whom have reflected thoughtfully about a society shaped by machines, megamachines, and systems.

THE SECOND DISCOVERY OF AMERICA

The historian of technology and science finds concepts and sees symbols of America and the American different from those of Turner and other highly influential American political, military, social, and economic historians. Instead of asking the reader to watch the procession through the Cumberland Gap, the historian of technology asks you to stand on the streets of industrial Chicago in 1900 and observe the skilled mechanic, the blue-collar worker, the eager young inventor, the aspiring white-collar manager, and the upwardly mobile, clean-shaven farmboy learning to be a mechanical engineer. They flourished in the turn-of-the-century decade when their country, once despised by industrial Britain as rural and uncouth, became the world's preeminent industrial and technological power. By 1900 the nation, having been a construction site for centuries, had spawned a nation of builders. The business of America was not business; it was building. The spirit of the people was not so much free enterprise as demiurge—the spirit of the Greek god who created the material world.

On occasion strangers see us more clearly than we see ourselves. Perceptive European visitors stood on the streets of our industrial cities early in this century and saw the procession of inventors, engineers, entrepreneurs, and workers that I have asked you to imagine. Like the historian of technology today, European visitors also saw America as essentially a building nation, a technological rather than a political and business-driven one. The displacement of the wilderness and the prairie by the machine especially impressed them. Visiting during the tumultuous half-century from 1880 to 1930, they discovered Edison introducing incandescent lighting in New York City; Philadelphia and Baltimore were scenes of mushrooming manufacturers of heavy chemicals, iron, and steel; and Boston was the site from which Bell's telephone networks spread. These, too, were the decades when Chicago showed the prodigious power of the industrial city by reversing the flow of the Chicago River to carry off the city's waste and when New York City coordinated her bridges, subways, and elevated and mainline railroads into the world's largest and most complex system of urban and regional transit. In both cities, the skyscrapers symbolized the constructive power of the industrial city. In Minneapolis, Minnesota, and Buffalo, New York, the stark forms of the great grain silos represented the might of mechanized agriculture. At roaring Niagara Falls, electrical engineers and chemists had built an integrated hydroelectric and industrial complex that became a model for technological development throughout the world. On the outskirts of Detroit rose the Ford River Rouge Plant, history's largest production machine.

Because the changes were concrete and visual, European artists and architects perceived them especially well. Visiting the great industrial metropolises, these new discoverers from Europe took the machine to be the symbol of the new America. Marcel Duchamp in 1915 scouted machine-made America for other avant-garde artists. Arriving in Manhattan, he proclaimed New York City a complete work of art. The only earlier works of art in America, he insisted, were its bridges and plumbing,[3] which reminds us that Duchamp wanted to display an ordinary urinal in a 1917 art show. Francis Picabia, a close friend of his, who arrived in the same year, said:

> This visit to America . . . has brought about a complete revolution in my methods of work. . . . Prior to leaving Europe I was engrossed in presenting psychological studies through the mediumship of forms which I created. Almost immediately upon coming to America it flashed on me that the genius of the modern world is in machinery and that through machinery art ought to find a most vivid expression. . . . I have been profoundly impressed by the vast mechanical development in America. The machine has become more than a mere adjunct of human life. It is really a part of human life—perhaps the very soul. In seeking forms through which to interpret ideas or by which to expose human characteristics I have come at length upon the form which appears most brilliantly plastic and fraught with symbolism. I have enlisted the machinery of the modern world and introduced it into my studio. . . . I don't know what possibilities may be in store. I mean simply to work on and on until I attain the pinnacle of mechanical symbolism.[4]

Earlier both Duchamp and Picabia had decided that all existing modes of art seemed inadequate for expressing modern concepts. In the machine, especially in its erotic significance and its demonstration of human creativity, they believed they had found their long-sought mode of expression.[5] In their stress on creativity in this man-made dynamic world, we find a reflection of the dramatic impact that the making of America was having on the war-torn, despairing European world.

Duchamp and others of the European avant-garde, like Sigfried Giedion later in his *Mechanization Takes Command,* believed that they could discover the unconscious of America in its machines and manufactures, especially its everyday artifacts. Duchamp pioneered in the effort to aestheticize the commonplace, to ennoble the banal—a motif extending through the work of Fernand Léger and exemplified more recently in Pop Art. These artists found in America the most interesting raw material for this cultural exploration, and they tried to awaken in technological America a self-awareness. They held up a mirror to modern Americans then;[6] the historian of technology and science captures the reflection now.

European artists saw in the American machine a symbol of modernity. Other Europeans also discovered the American system of production. The excited reaction of many Europeans to technological America, especially after World War I, has appropriately been called "the second discovery of America."[7] Escaping most of the violence and destruction of World War I, relatively free of the taint of contributing to its causes, and having displayed impressive industrial power and moral vigor, America enjoyed among Europeans, especially the defeated Germans, unprecedented prestige and popularity. Not surprisingly, therefore, they looked to American institutions and characteristics for sources of strength.

During the short life of the Weimar Republic, German political liberals, socialists, business people, and engineers also made the intellectual and ocean voyages of the second discovery. Germans looked to the source of American dollars that was shoring up their newly constituted and fragile republic and found there the mightiest of modern nations, a mode of production pouring forth forty-nine percent of the world's finished products. During the period of stabilization, German middle-class liberals, elite trade unionists, and other defenders of Weimar took American technological civilization as a model for imitation. Unlike so many other Americans then, they did not find the business of America business, but building. They envied America's high standard of living and its functioning democracy, but their curiosity about, and admiration of, America's highly rationalized production system commanded the focus of their attention. Many admirers decided that the interaction between technology and democracy was the essence of the nation. German liberals concluded that a German adaptation of the American combination of modern production technology and democracy could preserve the Weimar Republic. Revisionist socialists saw American large-scale, centralized production-technology as a step toward the socialist society.

Idealizing distant industrialized America like seventeenth-century Europeans had similarly idealized the American wilderness and the noble savage, Weimar Germans believed they had discovered a modern land in which there was peace between labor and capital, steadily rising wages, rapidly increasing consumption, and generally a hitherto undreamed of prosperity. Many took the United States to be the first classless society where everyone had an equal opportunity to own his or her own house and car and enjoy personal freedom. The sine qua non, these Germans presumed, was the organized, centralized, large-scale system of production. Taylorismus (scientific management) and Fordismus (mass production) constituted Amerikanismus.[8] The Europeans who made the great and second discovery of America, the nation of machines, megamachines, and systems, sensed what we now should know: the emergence of built, ordered, organized America was a historical event of surpassing importance.

TAYLORISMUS, FORDISMUS, AND AMERIKANISMUS

The second discoverers decided that Frederick W. Taylor and Henry Ford epitomized modern America, not J. P. Morgan, General John J. Pershing, or Calvin Coolidge. To these Europeans—and to the historian of modern American technology—the production system and the machine represented modern America. Even though both Taylor and Ford were deeply involved with capitalistic enterprise and the latter was an immensely wealthy capitalist, the Europeans thought of Taylor as a production engineer and of Ford as an industrial entrepreneur. Both can be characterized as system builders, as expressions of the American demiurge.

What is a system builder? In the case of Taylor, the father of scientific management, his single-minded objective was to integrate the workers into a machinelike system of production. He intended to do for the machine-worker interaction what nineteenth-century engineers had done for the interacting components of machines: to eliminate waste motion, save energy, and rationalize functional relationships. The worker would become a component in a more encompassing mechanism, or system, for production. Taylor wrote—and for this he became despised by many workers—that in the past people had been first but, in the future, the system must be first.

Henry Ford's plants at Highland Park and River Rouge testified to his prowess as a system builder. These mammoth industrial complexes functioned like a single machine, or system, composed of countless machines and processes functioning as subsystems. Visitors saw Ford's River Rouge as "a huge machine . . . each unit as a carefully designed gear which meshes with other gears and operates in synchronism with them, the whole forming one huge, perfectly-timed, smoothly operating industrial machine of almost unbelievable efficiency."[9]

Archilochus, the Greek poet, would surely have found Ford, Taylor, and other American system builders, including those who built the electric power and communication networks,[10] to be "hedgehogs" who know one big thing rather than foxes who know many things. Hedgehogs, as Sir Isaiah Berlin has added, "relate everything to a single central vision, one system less or more coherent."[11] The Greeks would also have associated these system builders with Demiurge, the creator of the material world and the husband of Necessitas.

MACHINES AND MEGAMACHINES

If the second discoverers of America and the historians of technology rightly stress the building of America and the technical and social order and control that follow from the creativity of such system builders as Ford and Taylor, then we should ask what special problems arise from the technological characteristics of built America. Not only have Americans in their celebration and sentimentalization of democratic politics and free enterprise overlooked the depth and extent of the mechanization and systematization of America, but they have also failed, with a few notable exceptions, to comprehend the deeper meaning of mechanization and systematization for their individual and social existences. Social historians have already told us of the degradation of the workplace and the alienation of mass production; today, social critics warn us of the pollution and destruction by technology of the natural environment; but rarely do we discuss and reflect on Frederick W. Taylor's dictum that in the past people had been first but, in the future, the system must be first. We celebrate Charles Darwin for discerning patterns in the natural world; we do not yet sufficiently appreciate the importance of finding patterns in the technological world.[12]

After the great burst of inventive and system-building activity during the decades from about 1880 to 1930, only a few historians and social critics began to assess the character of mechanized and systematized America. Among them, Lewis Mumford (b. 1895) and Sigfried Giedion (1888–1968) stand out because they emphasized the development and pervasive influence of modern machines, megamachines, and systems on the pattern, or structure, of individual lives and of society. Their reactions and diagnoses are worthy of reflection.

Giedion was a Swiss architectural and art historian who was able to contrast traditional and modern cultures. In 1948, following extended visits in the United States that had begun a decade earlier, Giedion published *Mechanization Takes Command*. In the foreword, he declared his intention of responding to a pressing problem that had arisen in the human condition. "By investigating one aspect of our life—mechanization," he wanted to show how a break between thought and feeling had occurred.[13] He insisted that society must reinstate basic human values by deciding how far mechanization corresponded with, and to what extent it contradicted, the laws of human nature. The onset of mechanization, he believed, split modes of thinking and modes of feeling.

The dramatic impact of his visits to America after 1938 helps explain

Giedion's apprehensions about the spread of mechanization. His fascination with mechanization may have become a fixation when he first tasted American bread. Afterwards, he found mechanization a part of the psychological and social fabric of American life. Giedion's concept of mechanization, however, is complex, if not elusive. At times mechanization is a process of production; at others it is its products. On other occasions, mechanization interests him as an effect that machine processes and products have on people. In *Mechanization Takes Command,* he chronicles, for instance, the transformation of the making of door locks by handicraft to their production by machines; he describes the development of assembly-line production in America, for he takes that to be the symptom of full mechanization. He devotes a long chapter to the American mechanization of the bath, for he found that this changed an act of leisure, sociability, and regeneration to one of efficient cleansing, a transformation representative of so many others occurring as mechanization took command.

Giedion insightfully associates twentieth-century mechanization with motion. To him, movement is one of the "Springs of Mechanization."[14] The nineteenth century was the first that learned "to feel the pulse of nature."[15] Giedion finds production-engineer Frank Gilbreth's (1868–1924) motion pictures of the work process, Paul Klee's drawings, and Marcel Duchamp's "Nude Descending a Staircase" (1912) revealing twentieth-century dynamism at a deep level. But it is the ceaseless flow of the moving assembly line, either disassembling animals for the meat packer or assembling autos for mobile Americans, that fixes the relationship between modern mechanization and motion.

Having noted Giedion's varied concepts of mechanization, we return to his leading question in *Mechanization Takes Command:* How has mechanization brought about a break between thought and feeling? He explores the problem in a concluding section entitled "Man in Equipoise."[16] Here he finds an analogy between mechanization and the forces of nature, even though mechanization springs from the human mind. Both forces are blind, he believes, and without direction of their own. Both can imperil the human condition or be applied to its advantage. The analogy is provocative but flawed, as Giedion inadvertently reveals in observing that mechanization comes from the human mind. This mind has given mechanization, unlike nature, human direction and purpose, even though these features may correspond only to those of the creators or designers of mechanization. Mumford, as we shall see, avoids this error in his definition of a megamachine for he finds it a rationalized means to the ends of those who design it, and he abhors, for instance, the megamachines of the authoritarian rulers and the militarists because they embody their objectives.

Having compared mechanization with natural forces, Giedion then attributes the growing loss of faith in progress in the twentieth century to the application of the forces of mechanization to unsuitable domains.[17] One of these is the bath. Mechanization in inappropriate domains leads to loss of psychological equilibrium. Here we find Giedion assuming that the human personality is dichotomized into the emotional and the rational and believing that these two aspects of the personality need appropriate environments and sustenance, respectively emotional and rational. The mechanized environment is appropriate for rational man, although it can destructively repress the emotional—hence, the danger of loss of equilibrium if mechanization takes command.

Giedion passionately pleads for a balance between rational and emotional human activities and between the mechanized environment and organic surroundings. He does not equate the organic only with nature. The mechanized environment can also have an organic, holistic quality that sustains the life of emotion. So Giedion calls for less narrow specialization, rationalization, and mechanization, and more generalization to achieve an organic synthesis of spirit and environment. He states that "we have refrained from taking a positive stand for or against mechanization. We cannot simply prove or disprove. One must discriminate between those spheres that are fit for mechanization and those that are not."[18] In his desire for the organic, Giedion is at one with Mumford.

MUMFORD AND THE MEGAMACHINE

For Mumford, the concept of mechanization did not encompass the essence of technology. He used the idea of a "megamachine" to express the profundity of his view of technology. The megamachine is not simply a physical machine; it is a rational and purposeful organization of human and other resources—a technological system. Mumford believed that the basic human inclination to order and control expresses itself in megamachines that take varied forms, ranging from the system of labor and tools that the pharaohs used to build the pyramids to the American military-industrial complex used to arm the nation. Mumford's megamachine results from the same rational, nonorganic drive that Giedion associates with mechanization, but, as noted, Mumford differs from Giedion in that the outcome is not simply a physical mechanism.

Mumford has written two major books about the history of technology, *Technics and Civilization* (1934) and *The Myth of the Machine: The Pentagon of Power* (1970). He defined the goal of *Technics and Civilization* as an exploration of the limitations and possible benefits of the machine to life, collective and individual. By the time he wrote *The Pentagon of Power,* he was "driven, by the wholesale miscarriages of megatechnics, to deal with the collective obsessions and compulsions that have misdirected our energies, and undermined our capacity to live full and spiritually satisfying lives."[19] In the process of writing the two books, Mumford transformed machine into megamachine. In *Technics,* the machine was becoming an embodiment of social and psychological attitudes.[20] In *Pentagon,* the concept had become explicit.

Throughout history people have purposefully and rationally ordered and controlled other people and things, according to Mumford, to solve prevailing social and psychological problems. During the early medieval era—the so-called Dark Ages—humans needed order in symbol and substance because of their helplessness in the face of natural forces and social chaos. In Europe during the eighteenth and nineteenth centuries factory, mill, and mine owners and the military monstrously distorted life and matter to enhance their power through the forces of production that they ordered and controlled.[21] In this period of rapid mechanization of production, Mumford, like Giedion, sees with dismay the subjugation of the vitalist or organic side of human nature by the mechanistic.

When writing *Technics,* Mumford, the social critic and reformer, saw the possibility that society would reverse the trend to mechanistic order and control, for then he did not realize the full dimension and pervasiveness of the megamachine. He was hopeful about the possibility of transcending the limiting characteristics of the machine and the mechanistic approach to life by accentuating those characteristics that allow the human spirit to flourish not only along rational lines, but also as a rich, diverse, complex, organic growth. This advance was possible, however, only if the control of technology changed hands. Technology under the control of the financiers, industrialists, and the military, as it had been since the eighteenth century in the West, could not contribute to the improvement of the human condition, because the vested interests of capitalism desired "to reduce all existence to terms of immediate profit and success."[22]

Writing in 1930, Mumford damned the profit seekers but idealized the scientists. He believed that first-rate scientists such as Michael Faraday and Willard Gibbs dedicated themselves to the life of science not only to master nature for the good of humanity, but also as a way of disciplining the mind and the spirit. He also concluded that physiology, a life science, had emerged in the late nineteenth century as the dominant science replacing mechanics. The machine would give way to organic technologies. Scientists had also become the creators of the new, nonmechanical technologies of the late nineteenth and twentieth centuries, electricity and chemistry. He was defining technology as applied science, as derivative of scientific laws, a position that few historians of technology would take today.

The good society would arrive when persons motivated like pure scientists used technology to create a "Green Republic." "We must turn society," he wrote early in the 1930s, "from its feverish preoccupation with money-making inventions, goods, profits, salesmanship, symbolic representations of wealth to the deliberate promotion of the more humane functions of life."[23] He believed that electric power would contribute greatly to the "Green Republic" for it would allow industry and the population to be dispersed into small green rural communities with power transmitted from distant power plants. No longer would industry need to concentrate near coal mines or waterpower sites. New modes of electric communication and transportation would also contribute to decentralization. In these small green communities, modern technology used by persons filled with a sense of craftsmanship would support the living of a balanced life filled with companionship, love, creativity, and a harmonious balancing of emotional and physical needs.

Recognizing the possibility that survivals of an earlier industrial era could possibly destroy the emerging society and technology, Mumford qualified his optimism in the 1930s.[24] Yet the reader of *Technics and Civilization* could hardly have anticipated the profound pessimism of Mumford almost forty years later, when he wrote *The Pentagon of Power*. He no longer envisioned a humanistic technology and control of technology passing to people free of avarice and anger. By 1970 the dimensions of "the machine" had become frighteningly clear to him as the "megamachine," a horrendous organization of power that mechanized society. The megamachine imposed purely mechanical forms on every manifestation of life, "thereby suppressing many of the most essential characteristics of organisms, personalities and human communities."[25] The proponents of the megamachine denied life by reducing it to the quantitative, mechanical, and chemical.

No longer did Mumford expect scientists to humanize technology: "to become the 'lords and possessors of nature' was the ambition that secretly united the conquistador, the merchant adventurer and banker, the industrialist, and the scientist, radically different though their vocations and their purposes might seem."[26] Science was becoming increasingly irrelevant to human intentions except those of the corporate enterprise or the military establishment. Scientists no longer simply sought order in the universe, and reported what they had found; rather, they selected those aspects of nature that could be organized systematically into a mechanistic world view. Scientists were now confusing what they thought to be their higher order of reality with a sterile higher order of abstraction.[27] Galileo Galilei became the symbol for Mumford of the many scientists who had transformed a complex world—Baroque, for Galileo—into a quantitative, objective, sterile wasteland. He wrote of "the crime of Galileo."

Mumford's pessimism after World War II can be attributed in large part to the dropping of the bombs on Hiroshima and Nagasaki. That news in 1945 so stunned him that discourse became impossible for him days afterward.[28] The mammoth organization needed to make the bombs was sufficient evidence of the spread of the megamachine, and the involvement of the physicists in the project deprived him of the hope that scientists would preside over the making of a new technological society less influenced by the drive for power and wealth.

ELLUL'S TECHNOLOGICAL SYSTEMS

Giedion expressed concern about the displacement of the organic and emotional aspects of life by the mechanistic; Mumford, more deeply troubled, wrote of the megamachine suppressing many of the most essential characteristics of humans and their communities. Jacques Ellul, the French philosopher and cultural critic, has also explored the implications of the megamachine, or technological system, and found the "technological system" a dire threat to human freedom.[29] Americans, long proud of nurturing and defending political freedom, should heed especially Ellul's warning that political activity is unreality; technology is reality.[30] He finds the technological system that has swallowed up the capitalistic system a far greater threat to our freedom of action than authoritarian politics. The order, method, neutrality, organization, and efficiency of the technological system has transformed humankind into a technicized component of the technological system, a component devoid of will, analogous to one in a machine or megamachine.[31]

Ellul declares that technology changes the natural and cultural environments by fragmenting their realities and then recombining the discontinuous fragments into technological systems that are operational, instrumental, and organized for problem solving. The old complexity of culture and nature is replaced by a reductionist technological one. Neither nature nor culture now determines social structure. The technological system has become the determiner.[32] Denying the primacy of politics, Ellul argues that the state is no longer as influential a factor as technological systems. The threat to human autonomy is no longer as great from government as from technology.[33]

Young people today, he believes, are not liberally educated but trained to function in technological systems; not being liberally educated, they have no basis from which to criticize the systems that embrace and direct them.[34] Today we can no more dream of challenging the technological milieu than a twelfth-century man would dream of objecting to trees, rain, a waterfall. Humans, he adds, have "no intellectual, moral, or spiritual reference point for judging and criticizing technology."[35]

CONCLUSION

I have argued in this essay that historians attentive mostly to political, economic, and social experiences have shaped our memories and concepts of our national characteristics. As a result, Americans tend to sustain an ideology that values highly the American democratic spirit and free business enterprise. In contrast, the historian interested in technology presents other images of the past and finds the preeminent American characteristic to be the commitment to building, to creating a technological world. The testimony of Europeans who visited the United States during the early twentieth century, after the country had achieved preeminence as an industrial and technological nation, reinforces the interpretation of the historian of technology. The Europeans, especially Germans, eager to discern the essence of the American nation, made a "second discovery of America," the discovery of a nation the character of which was best symbolized by machines, megamachines, and systems, not only legislative bodies and business corporations.

If the nation, then, is essentially a technological one characterized by a creative spirit manifesting itself in the building of a technological world patterned by machines, megamachines, and systems, then we should heed those such as Lewis Mumford, Sigfried Giedion, and Jacques Ellul, who fathom the depths of the technological society, who identify currents running more deeply than those conventionally associated with politics and economics. If we follow their lead and accept the interpretation of the historian of technology, then we shall realize that our deep and persistent problems arise from the mechanization of life; the sacrifice of the organic in the name of order, efficiency, and control; and the loss of freedom because of the constraints placed on us by the host of large technological systems that structure our world. The forces that we should understand and control in order to shape our destiny are not now primarily natural or even political, but technological.

For example, the recent shuttle tragedy revealed, in the network of responsibility and action, the complex array of astronauts, O rings, engineers, managers, government departments, and industrial corporations involved in the NASA system. Similarly, the Chernobyl disaster showed that the Soviet nuclear power plants and electrical supply network are parts of a large and complex technological system that includes political and economic institutions organized to supply power for industrial and other purposes. Societal values calling for frequent, economic, and newsworthy launchings that garner public support, in the case of NASA, and values calling for a rapid increase of low-cost energy in the name of national achievement, in the case of the Soviet energy system, are integral parts of these technological systems.

American schools struggle now to teach how the American political system works and how the citizen may exercise some control over it rather than be subject to it. This pedagogic and civic enterprise is commendable. Yet, if we attend, among others, the second discoverers of America, social critics like Mumford, and the historians of technology, then we should realize that we face pedagogical and civic problems of greater complexity than the conventional political ones. It is important to understand how Congress works, but recent events suggest that it may be as—if not more—pressing for the citizen, and those who inform and instruct the public, to understand how technological systems, broadly conceived to include institutions and values, work. The purpose of the understanding is not simply to comprehend the impressively ordered, controlled, and systematized, but to exercise the civic responsibility of shaping those forces that in turn shape our lives so intimately, deeply, and lastingly.

NOTES

I wish to acknowledge the assistance of Karl Michelson and Jane Morley, University of Pennsylvania, in preparing this essay.

1. The literary critic, Perry Miller, has provided a marvelous image of nineteenth-century Americans, an image vastly different from Frederick Jackson Turner's, derived from the frontier. Miller imagined the passion with which the Americans "flung themselves into the technological torrent, how they shouted with glee in the midst of the cataract, and cried to each other as they went headlong down the chute that here was their destiny." See Perry Miller, "The Responsibility of Mind in a Civilization of Machines," *American Scholar* 31 (Winter 1961–62): 55.

2. Frederick Jackson Turner, "The Significance of the Frontier in American History," originally presented at the Ninth Annual Meeting of the American Historical Association, 11–13 July 1893, Chicago, and published in the Association's *Annual Report for the Year 1893* (Washington, D.C.: Government Printing Office, 1893), p. 227.

3. Stanislaus von Moos, "Die Zweite Entdeckung Amerikas," in Sigfried Giedion, *Die Herrschaft der Mechanisierung: ein Beitrag zur anonymen Geshichte,* with an afterword by Stanislaus von Moos (Europäische Verlagsanstalt) (Frankfurt am Main, 1982), p. 807. Giedion's *Mechanization Takes Command: A Contribution to Anonymous History* (New York: Oxford University Press), was only recently translated into German.

4. Frederick Macmonnies, "French Artists Spur on an American Art," *New York Tribune,* 24 October 1915, IV, p. 2. Quoted by William B. Canfield, "The Machinist Style of Francis Picabia," *Art Bulletin* 48 (1966): 309, 313, and cited in *The Machine as Seen at the End of the Machine Age,* ed. K. G. Pontus Hultén (catalog of the exhibition, The Museum of Modern Art, New York, 25 November–9 February 1969) (New York: Museum of Modern Art, 1968), p. 83.

5. Ibid., p. 82.

6. Von Moos in *Herrschaft,* pp. 806–7.

7. Ibid., p. 781.

8. Jost Hermand and Frank Trommler, *Die Kultur der Weimarer Republik* (Munich: Nymphenburger Verlagshandlung, 1978), pp. 49–58.

9. Van Deventer as quoted in Allan Nevins and Frank Ernest Hill, *Ford,* vol. 2: *Expansion and Challenge, 1915–1933* (New York: Scribners, 1957), p. 288.

10. On the evolution of electric power systems and the system builders, see Thomas P. Hughes, *Networks of Power: Electrification in Western Society, 1880–1930* (Baltimore: Johns Hopkins University Press, 1983).

11. Isaiah Berlin, *The Hedgehog and the Fox: An Essay on Tolstoy's View of History* (New York: Simon and Schuster, 1953), p. 1.

12. Thomas P. Hughes, "The Order of the Technological World," *History of Technology* 5 (1980): 1–16.

13. Giedion, *Mechanization Takes Command,* p. v.

14. Ibid., pp. 13–14.

15. Ibid., p. 17.

16. Ibid., pp. 714–23.

17. Ibid., p. 716.

18. Ibid., pp. 720 (quote), 721.

19. Lewis Mumford, *The Myth of the Machine: The Pentagon of Power* (New York: Harcourt, Brace, Jovanovich, 1970), p. 1.

20. Lewis Mumford, *Technics and Civilization* (New York: Harcourt, Brace, 1934), p. 3.

21. Ibid., pp. 4–5.

22. Ibid., p. 219.

23. *The Nation* 133 (9 December 1931): 631.

24. Mumford, *Technics,* p. 214.

25. Mumford, *Pentagon of Power,* p. 37.

26. Ibid., p. 78.

27. Ibid., p. 74.

28. Author's conversation with Lewis Mumford and Sophia Wittenberg Mumford in Amenia, New York, 28 December 1985.

29. Jacques Ellul, *The Technological System,* trans. Joachim Neugroschel (New York: Continuum, 1980).

30. Ibid., p. 16.

31. Ibid., p. 7.

32. Ellul can be read as a technological determinist, but not Mumford. Culture in the beginning, Mumford believes, did not follow from the use of tools. He also denies technological determinism in the modern era. Like so many historians and sociologists of technology today, he believes that technology is a cultural artifact, but one that reacts back on and shapes the culture that made it. Giedion's position on technological determinism is neutralist (*Mechanization,* p. 345). He argues on one occasion that the tool user—society—determines the effects of the technology, but he also writes that mechanization resulted from a rationalistic view of the world, thus suggesting cultural or psychological determinism (*Mechanization,* pp. 3, 32).

33. Ellul, *System,* pp. 56–57, 311.

34. Ibid., pp. 45–46, 48, 312–13.

35. Ibid., p. 318.

History and Philosophy of Technology: Tensions and Complementarities

PAUL T. DURBIN

A built-in tension exists between abstraction-oriented philosophers of technology and detail-oriented historians of technology, but the two fields also have a complementary character.

Interestingly, Melvin Kranzberg, so instrumental in establishing the Society for the History of Technology (SHOT), was also influential in the movement to establish the Society for Philosophy & Technology (SP&T). Showing his familiarity with that society—and with the churlish habits of philosophers—he wrote to me in 1979, when SP&T was starting to elect officers and do the other things usually associated with an academic professional society: "I think that SPAT would be a wonderful acronym for your Society. Easy to say, and also very descriptive of the way in which philosophers treat one another." In 1986, Kranzberg wrote to me again: "It has given me a warm feeling to see the development of scholarship in the field of the philosophy of technology and the growth of the Society for Philosophy & Technology. I hope that I can be of some help in the continuing growth of this very important field."

Kranzberg had certainly been helpful up to that point. He was centrally involved in setting up a SHOT symposium on philosophy of technology—eventually published in *Technology and Culture* (Summer 1966)—that included contributions by philosophers Joseph Agassi, Mario Bunge, and Henryk Skolimowski and by historian Lewis Mumford. In 1973, Kranzberg published Carl Mitcham and Robert Mackey's pioneering *Bibliography of the Philosophy of Technology* as a special number of *Technology and Culture*. In 1975, he provided the names of several key philosophers to be invited to the first University of Delaware conference on philosophy and technology—the source of most of the papers that appear in volume one of *Research in Philosophy & Technology,* the official publication of the Society for Philosophy & Technology from 1978 until 1985. Kranzberg served on the editorial board of that publication and also serves on the editorial board of the society's new series, *Philosophy and Technology.*

Why would a historian of Kranzberg's broad influence be interested in encouraging philosophers to begin to study technology? In a review of the book *Philosophy and Technology,* Andrew Lugg has written: "The trouble is that technology is all too often marshaled to illustrate philosophical doctrine rather than the other way around. Surely what is required is a careful analysis of specific technologies, not more reflection on Technology understood in the broadest possible sense."[1] Is it not true that philosophers of technology have dwelt too long on Technology and given too little of their attention to the "careful analysis of particular technologies"? Even in philosophical circles, no one thought it odd when, as recently as 1976, a Philosophy of Science Association symposium was labeled, "Are There Any Philosophically Interesting Questions in Technology?" Symposiasts included Max Black, Mario Bunge, Edwin Layton, and myself, all giving affirmative answers. But there was also a reply by Ronald Giere, arguing that there was at that time still no recognizable paradigm for philosophy of technology.[2]

TENSIONS BETWEEN PHILOSOPHY AND HISTORY OF TECHNOLOGY

From earliest times in the history of Greek philosophy, it was a popular sport to poke fun at or otherwise disparage philosophers' preoccupation with celestial rather than down-to-earth phenomena, their passion for the obvious, their compulsion to deal (only) with foundations. All the same, at least from the time of Socrates on, the most civilized Western societies have always recognized that they need such head-in-the-clouds thinkers. In contrast, there can be no worse criticism of a historian than that he or she has (merely?) "discovered" the facts suggested by some preconceived theory or prejudice. John Staudenmaier calls historians of technology "technology's storytellers."[3] In those terms, most people would agree that philosophers are the distinctionmakers, the clarifiers, the systematizers of the stories that others tell.[4] As a devotee of American Pragmatism, I believe that philosophers should also act in the context of the problem-solving efforts of the larger community.

There are obvious tensions within the disciplines of history and philosophy, between the two, and with the larger society that supports both. I believe, however, that some critics make too much of them. In what follows, I concentrate instead on what historians of technology and philosophers of technology can contribute to one another's efforts.

THE GROWTH OF PHILOSOPHY OF TECHNOLOGY

When SHOT invited a group of philosophers to explore the possibilities of a philosophy of technology, in the symposium that appeared in *Technology and Culture* in 1966, there was already a limited body of literature available on the philosophy of technology. In one of his careful historical surveys of the development of the field, Carl Mitcham mentions earlier works in Germany: Ernst Kapp's *Grundlinien einer Philosophie der Technik* (1877), Friedrich Dessauer's *Philosophie der Technik* (1927) and *Streit um die Technik* (1956), and a series of VDI (Association of German Engineers) conferences responding to the "culture critics" of technology, especially Martin Heidegger.[5] Mitcham also notes important work in Eastern Europe, tracing its roots to Karl Marx (who probably deserves to be called the first modern philosopher of technology) and with an abundant literature after World War II (earlier than but culminating in *Man, Science, Technology: A Marxist Analysis of the Scientific and Technological Revolution*).[6] For the United States, Mitcham cites Lewis Mumford's *Technics and Civilization* (1934) as an early forerunner and Mumford's two-volume *The Myth of the Machine* (1967, 1970).[7] Herbert Marcuse's best-known work, *One-Dimensional Man*, was published just before the *Technology and Culture* symposium, as was the English translation of Jacques Ellul's *The Technological Society*.[8] As Mitcham also notes, there had been, from the very beginning of artificial intelligence (AI) research, a great deal of interest by philosophers in that topic.

After the *Technology and Culture* symposium, there was a burst of activity mainly involving philosophers with appointments in technical institutes (with a few at regular universities). An anthology, *Philosophy and Technology*, was edited by Carl Mitcham and Robert Mackey in 1972. In 1973, Mitcham and Mackey published their bibliography, first as a special number of *Technology and Culture*, then in book form. That same year, the Hastings Center's Daniel Callahan published *The Tyranny of Survival*, a book that stresses, among other things, the strong technological impact on contemporary biomedicine. Friedrich Rapp's anthology, *Contributions to a Philosophy of Technology: The Structure of Thinking in the Technological Sciences*, appeared in 1974.[9]

In 1975, the University of Delaware hosted the first conference on philosophy and technology, and a newsletter was begun. The *Newsletter of the Society for Philosophy & Technology* is still being published; the current editor is Edmund Byrne, Indiana University at Indianapolis. In 1976 the Philosophy of Science Association took sufficient note to hold the previously mentioned invited-paper symposium on whether philosophy of technology had anything interesting to offer. Also that year engineer Samuel Florman's *The Existential Pleasures of Engineering* appeared.[10] The first two books by American philosophers of technology appeared in 1977: Bernard Gendron's *Technology and the Human Condition*, and Langdon Winner's *Antonomous*

Technology.[11] In 1978, *Research in Philosophy & Technology,* volume 1, appeared, as did Rapp's *Analytische Technikphilosophie* (English translation, *Analytical Philosophy of Technology,* 1981).[12] In 1979, Don Ihde's *Technics and Praxis* appeared in the U.S. and Hans Jonas's *Das Prinzip Verantwortung* in Germany (English translation, *The Imperative of Responsibility: In Search of an Ethics for the Technological Age,* 1984).[13]

In the 1980s, activity has continued at a brisk pace. Mitcham's survey, "Philosophy of Technology," took its place alongside other disciplinary surveys in *A Guide to the Culture of Science, Technology, and Medicine.*[14] (Melvin Kranzberg was area editor for the history of technology chapter written by Carroll Pursell.) That same year, the first detailed study of a particular area of technology by an American philosopher of technology appeared—Kristin Shrader-Frechette's *Nuclear Power and Public Policy.*[15]

The first international conference on philosophy and technology, involving North Americans and West Germans, was held in April 1981 in Bad Homburg, West Germany, with support from the U.S. National Science Foundation and its West German counterpart. Proceedings of the conference were published in German, *Technikphilosophie in der Diskussion,* and in English, *Philosophy and Technology.*[16] Also in 1981, Fred Dretske published *Knowledge and the Flow of Information,*[17] continuing American philosophers' interest in AI research and related topics. Friedrich Rapp's survey (with an excellent international bibliography), "Philosophy of Technology," appeared in a survey of contemporary philosophy in 1982.[18] In 1983, the second international conference, with official backing of the Society for Philosophy & Technology, was hosted by Polytechnic University in New York City. The proceedings volume, *Philosophy and Technology II: Information Technology and Computers in Theory and Practice,* appeared in 1986.[19] In 1984, *Research in Philosophy & Technology* published its first theme volume, on technology and society, to coincide with "the Orwell year."[20] In 1985 the third international conference of the Society for Philosophy & Technology was hosted by Twente Technological University in the Netherlands. The topic was Technology and Responsibility.

Since 1980, so many books have been published in the broad area of philosophy and technology that even a sample list would have to include nearly a dozen items.[21] Anyone familiar with this body of literature might conclude that most philosophers who have discussed it are opposed to technology. This perhaps-too-common impression is superficial and misleading—probably based on the notoriety of such figures as Martin Heidegger and Jacques Ellul (who, admittedly, have gained a wide following in certain intellectual circles). However, as Carl Mitcham has emphasized, there is a fundamental tension among philosophers who deal with technology—between those who are basically protechnology and those who are critical—and discussions of both sorts appear in the literature in approximately equal percentages.

HOW PHILOSOPHY COMPLEMENTS THE HISTORY OF TECHNOLOGY

Methodological Similarities and Contrasts

I want to begin my list of complementarities by contrasting John Staudenmaier's history of *Technology and Culture* (and, more generally of SHOT), *Technology's Storytellers* (1985), with the work of Carl Mitcham on the history of the philosophy of technology. Staudenmaier maintains that in the short period of history he studies, mainly 1959 to 1980, there was a gradual shift among *Technology and Culture* authors from an early preoccupation with internalist history to a later and more mature stage focused on what he calls "the design-ambient contextual approach." He concludes with this bold claim:

> It is the argument of this entire book that historians of technology must choose between the ideology of progress [especially congenial to internalist historians of science and technology] and the design-ambient contextual approach as the governing model for their research. Both perspectives provide an encompassing model with which more specific studies can be situated, but the differences between the two are radical. It is impossible to adhere to the ideology of progress and write contextual history; the two views are incompatible to the core.[22]

Some historians of technology might claim that Staudenmaier is exaggerating, and they might trace the source of his bias to philosophy. Indeed, Staudenmaier himself acknowledges that he has a "general epistemological philosophy" derived from philosophers George Klubertanz (a fellow Jesuit) and James Collins, and also a "hermeneutical approach" borrowed from Hans-Georg Gadamer's *Truth and Method* (1975).[23] While historians critical of Staudenmaier's approach might use these acknowledgments to discredit his approach, I think they show the complementarity of history and philosophy of technology. History that is not a mere jumble of facts must make some assumptions, some of which will perforce be general enough to be called philosophical. Purportedly value-free research simply masks particular value choices.

Carl Mitcham has done at least two useful surveys of the historical background of philosophy of technology that can serve to reinforce this complementarity.[24] In the earlier formulation, Mitcham speaks of two usages of the phrase, "philosophy of technology," and claims that, in one sense, the phrase "has come to refer especially to engineering attempts to defend the profession against the hostility of the Romantic tradition."[25] In the later formulation, Mitcham elaborates:

> The "philosophy *of* technology" can mean two quite different things. When "*of* technology" is taken as a subjective genitive, indicating that

which is the subject or agent, philosophy of technology is an attempt by technologists or engineers to elaborate a technological philosophy. When "*of* technology" is taken as an objective genitive, indicating the object being dealt with, then philosophy of technology refers to an effort by philosophers to take technology seriously as a theme for systematic reflection. The first . . . tends to be more pro-technology, the second [at least] somewhat critical.[26]

Mitcham is clearly exaggerating for effect here, because a number of philosophers, as well as engineers and spokespersons for technology, have taken protechnology stances and claimed to offer a "serious" technological philosophy.[27] But Mitcham is making a serious point in the overly sharp distinction, which is evident in the eloquent conclusion of his first survey:

> The issue [of humanization] . . . is at the heart of philosophy of technology as a hermeneutic enterprise; it may also be at the heart of philosophy in general. . . .
> The deepest question to be addressed . . . concerns the place of philosophy itself in the technological world. . . . Two centuries of philosophical criticism, comparable in character to the centuries of criticism that ushered in the modern period, have been ineffectual in altering the course of modern commitments [to technological "progress"] in any fundamental way. . . . In reflecting on technology, philosophy is compelled to grapple with its own nature and the extent of its powers in ways that will, ironically, influence its analysis of technology.[28]

Staudenmaier is merely attacking "Whig history" and "the ideology of progress" among historians of technology; Mitcham would say that Staudenmaier's appeal to Gadamer's hermeneutic approach makes him part of a much larger group of intellectuals critical of the modern commitment to progress through applications of science and technology to the solution of social problems. Interestingly, just before his eloquent peroration, Mitcham adds the following: "In attempting to explore this [humanizing, hermeneutic] argument, philosophers make inevitable appeals to history." And they should do so. But, equally, "contextualist" historians of technology ought to be aware of the larger stakes at issue in their enterprise.

History of Technology Explicitly Addressing a Philosophical Claim

My example here is Thomas Hughes's *Networks of Power* (1983). Without naming Jacques Ellul or Langdon Winner or any other "autonomous technology" theorist, Hughes sets out in that volume (at least in part) to contrast the theory of technological autonomy with the complex reality of the development of the systems for electric power generation and distribution in Germany, Great Britain, and the United States in the period from 1880 to 1930. Hughes concludes:

Throughout this study it has been necessary to reach out beyond the technology, outside the history of technical things, to explain the style of the various systems, so obviously the technological systems were not simplistically autonomous, or free of the influence of nontechnical factors. The evolving power systems were not, metaphorically speaking, driverless vehicles carrying society to destinations unknown and perhaps undesired.

To protect himself from a charge of oversimplification, however, Hughes felt the need to add immediately: "The systems did, however, have an internal drive and an increasing momentum. The continuous emergence of reverse salients and the ongoing solution of critical problems by inventors, engineers, and entrepreneurs provided this."[29]

There are two difficulties with this conclusion—however massively documented it may be. Jacques Ellul could easily challenge a subsequent statement by Hughes: "If a would-be Darwin of the technological world is looking for laws analogous to the environmental forces that operate in the world of natural selection, the economic principles of load factor and economic mix are likely candidates." Do we not here have the unchallenged goal of Technique, economic efficiency? Ellul could say that Hughes's study, instead of refuting it, massively demonstrates the autonomy of technology.[30]

Less tendentiously, Langdon Winner might well challenge Hughes's assumption that "autonomous technology" can be equated with the metaphor of systems as "driverless vehicles carrying society to destinations unknown and perhaps undesired." What Winner would say, rather, is that technological systems have a power—partially borrowed from the ideology of progress, but also based on historical circumstances—to beguile their directors, and often whole societies, into wholesale acceptance without consideration for the social and political consequences. That social and cultural factors, and even idiosyncratic managerial decisions, move technological systems in particular directions is not evidence against the autonomous technology thesis; it indicates only that these social and cultural factors, and individual decisions, can operate without the benefit of wise, democratic political discussion and debate. What Winner would ask of Hughes is this: Was probing political discussion (at one point he likens such discussion to the debate engaged in by the founders of the United States) brought to bear on the development of these electric power systems? Or were managers and politicians able to persuade the public, without debate, of the inevitably progressive character of electric power?[31]

I do not intend to take sides on these issues. Hughes has not provided a devastating refutation of the autonomous technology thesis; however, Ellul's version of the thesis seems empirically unverifiable or nonfalsifiable. A careful reading of both provides a useful perspective.

History as Evidence for a Philosophical Theory

In *America by Design* (1977),[32] David Noble is not at all reticent in stating his Marxist slant: his book ends with a revealing (and persuasive) reference to the inevitability of class struggle even in the United States after generations of managerially manipulated social "reforms." Clearly, excellent history—no matter that purists can find fault with small details—can be written in support of a philosophical viewpoint. Equally revealing is the fact that many of the same points were being made at the same time, by way of sustained philosophical argumentation, in Bernard Gendron's *Technology and the Human Condition* (1977).[33] Thus, in this case (and probably also in others), history and philosophy of technology nicely balance and complement one another.

Philosophy of Technology Need Not Always Ignore Particular Technologies

Here we need to recall Andrew Lugg's observation: "What is required is a careful analysis of specific technologies, not more reflection on Technology."[34] Philosophers themselves have been sensitive to this charge, and at the end of the first international conference on philosophy and technology it was decided that the next conference should be devoted to a concrete technology. The result was *Philosophy and Technology II: Information Technology and Computers in Theory and Practice* (1986).[35] Although historians might claim that this volume remains at the abstract level, it shows that philosophers can deal with particular technologies. An even older, and better, example is philosophical treatments of artificial intelligence.[36]

History and Philosophy of Technology as Necessary Complements

I am currently doing research within the general philosophical framework of American Pragmatism that deals in different ways with the relationship, in countries with a high-technology-driven economic system, between the research-and-development (R&D) community and such traditional cultural institutions as museums, orchestras, universities, schools, the media, recreational institutions such as nature centers, and institutional religion. My particular interest is in social problems that have arisen (or that persist despite expectations to the contrary) in high-technology societies and in what the R&D institutions of modern society can (reasonably) be expected to do about them. It is my fundamental belief that such institutions have a moral, social, and even political responsibility to respond.[37]

In this sketch of complementarities between history of technology and philosophy of technology, I would not want to impose such a reform agenda on historians. However, to make my case credible, I need evidence from solid, reliable (and limited) history of technology studies.

Because I live and work in Delaware, a state with a high-technology base, I decided to explore the relationship between Delaware's R&D community (including the local high-tech research centers and their parent corporations) and the state's traditional cultural institutions. Can any generalizable lessons be learned from our experiences in Delaware? On this small scale, is it possible to say whether the growth of the R&D community since World War II has been, on balance, good or bad for the community—and good or bad especially in terms of the values usually associated with traditional cultural institutions? Although my investigations continue, it is clear that I would never be able to draw any worthwhile conclusions without relying on studies of the local research community done by colleagues in the history department at the University of Delaware.

Carol Hoffecker, in *Corporate Capital: Wilmington in the Twentieth Century,* documents the diminishing control that the du Pont family has had over the arts, society, and politics in the period after the Du Pont R&D workforce began to expand rapidly in Delaware.[38] Hoffecker shows how the corporation-related middle classes in the suburbs were at times able to withstand pressure from the du Pont family (e.g., on the issue of metropolitan government). But she also shows how this suburban population—so often made up of chemists and engineers and corporate middle management—was instrumental, by its indifference, in the deterioration of the city of Wilmington, and particularly in the worsening plight of its black citizens.

John Munroe, in *The University of Delaware: A History,* is much more defensive than Hoffecker with respect to that institution and its du Pont family and Du Pont Company benefactors.[39] However, even Munroe is forced to admit that the university's administration was heavy-handed in protecting its public image, and showed significant lack of respect for the academic freedom of its professors, particularly in the period of rapid build-up, after World War II, of research programs, notably in chemistry, chemical engineering, and other technical fields.

Most closely related to my work is a book by David Hounshell and John K. Smith, *Science and Corporate Strategy: Du Pont R&D, 1902–1980.*[40] Hounshell and Smith, among other things, document a number of significant instances in which the company barred, or at least delayed, the publication of important scientific work on proprietary grounds. They also show how Du Pont scientists fought against this policy, so much at odds with proper scientific procedures, with varying success at different times.

These three examples do not, in themselves, document a history of major social problems of technological society, but they represent instances of corporate practices that concern liberal reformers. Moreover, they do serve to illustrate the complementarity between history and philosophy of technology. Such philosophical work almost certainly requires collaboration with historians of technology, along with historians and sociologists of science and others. This approach fulfills the hopes of John Dewey, who called for teams of academics working with open-minded elected officials and other public leaders to collaborate on the solution of social problems using "creative intelligence."[41]

CONCLUSION

I realize that what I have said here remains open to the charge so often made by historians against philosophers—that I have dealt almost exclusively with generalities, or that I am merely restating the obvious. I take this risk, however, because I think that in this way I can draw some important lessons. And the main lesson I would draw, especially from my list of sample complementarities, is aimed directly at the audience for this volume.

Historians of technology, in my view, ought to become more aware of their philosophical presuppositions—either by reading more philosophy or by collaborating on projects with philosophers. I readily admit, at the same time and as a matter of course, that philosophers of technology (and philosophers of other disciplines) ought also to work harder at getting their facts straight and at being more honest about the relation between their theories and alleged factual support. But that amounts simply to repeating that history of technology and philosophy of technology complement one another—as Melvin Kranzberg has for so long wisely recognized. Simple claim or not, however, it has an important social consequence. If our joint efforts are to bear any fruit, we must work to break down, wherever possible, disciplinary barriers that still stand in the way of fruitful collaboration.

NOTES

1. Andrew Lugg, review of *Philosophy and Technology,* eds. Paul T. Durbin and Friedrich Rapp. *Isis* 76 (June 1985): 261.
2. See Frederick Suppe and Peter Asquith, eds., *PSA 1976* (East Lansing, Mich.: Philosophy of Science Association, 1977), 2:139–204.
3. John M. Staudenmaier, S. J., *Technology's Storytellers: Reweaving the Human Fabric* (Cambridge, Mass.: MIT Press, 1985).
4. The nature of philosophy in relation to philosophy of technology is discussed by Steven L. Goldman, "The *Technē* of Philosophy and the Philosophy of Technology," in *Research in Philosophy & Technology,* ed. P. Durbin (Greenwich, Conn: JAI Press, 1984), 7:115–44.
5. See Carl Mitcham, "Philosophy of Technology," in *A Guide to the Culture of Science, Technology, and Medicine,* ed. P. Durbin (New York: Free Press, 1980; paper, 1984), pp. 282–363; Ernst Kapp, *Grundlinien einer Philosophie der Technik: Zur Entstehungsgeschichte der Cultur aus neuen Gesichtspunkt* (Braunschweig, 1977; reprinted with new introduction, Düsseldorf: Janssen, 1978); Friedrich Dessauer, *Philosophie der Technik: Das Problem der Realisierung* (Bonn: F. Cohen,

1927), and idem, *Streit um die Technik* (Frankfurt: J. Knecht, 1956); Alois Huning, "Philosophy of Technology and the Verein Deutscher Ingenieure," in *Research in Philosophy & Technology*, ed. P. Durbin (Greenwich, Conn.: JAI Press, 1979), 2:265–71; Martin Heidegger, *The Question Concerning Technology and Other Essays* (New York: Harper and Row, 1977; lead essay originally published in German, 1954).

6. *Man, Science, Technology: A Marxist Analysis of the Scientific and Technological Revolution* (Moscow and Prague: Academia, 1973).

7. Lewis Mumford, *Technics and Civilization* (New York: Harcourt Brace, 1934); idem, *The Myth of the Machine*, vol. 1: *Technics and Human Development* (New York: Harcourt Brace Jovanovich, 1967), vol. 2: *The Pentagon of Power* (New York: Harcourt Brace Jovanovich, 1970).

8. Herbert Marcuse, *One-Dimensional Man: Studies in the Ideology of Advanced Industrial Society* (Boston: Beacon, 1964); Jacques Ellul, *The Technological Society* (New York: Alfred A. Knopf, 1964; French original, 1954).

9. Carl Mitcham and Robert Mackey, eds., *Philosophy and Technology: Readings in the Philosophical Problems of Technology*, 2d ed. (New York: Free Press, 1983; original ed., 1972); Carl Mitcham and Robert Mackey, *Bibliography of the Philosophy of Technology* (Chicago: University of Chicago Press, 1973); Daniel Callahan, *The Tyranny of Survival: And Other Pathologies of Civilized Life* (New York: Macmillan, 1973); Friedrich Rapp, ed., *Contributions to a Philosophy of Technology: The Structure of Thinking in the Technological Sciences* (Dordrecht: Reidel, 1974).

10. Samuel C. Florman, *The Existential Pleasures of Engineering* (New York: St. Martin's, 1976).

11. Bernard Gendron, *Technology and the Human Condition* (New York: St. Martin's, 1977); Langdon Winner, *Autonomous Technology: Technics-out-of-Control as a Theme in Political Thought* (Cambridge, Mass.: MIT Press, 1977).

12. Paul T. Durbin, ed., *Research in Philosophy & Technology*, vol. 1 (Greenwich, Conn.: JAI Press, 1978); Friedrich Rapp, *Analytical Philosophy of Technology*, Boston Studies in the Philosophy of Science, vol. 63 (Dordrecht: Reidel, 1981; German original, 1978).

13. Don Ihde, *Technics and Praxis*, Boston Studies in the Philosophy of Science, vol. 24 (Dordrecht: Reidel, 1979); Hans Jonas, *The Imperative of Responsibility: In Search of an Ethics for the Technological Age* (Chicago: University of Chicago Press, 1984; German original, 1979).

14. Paul T. Durbin, ed., *A Guide to the Culture of Science, Technology, and Medicine* (New York: Free Press, 1980; paper 1984).

15. Kristin S. Shrader-Frechette, *Nuclear Power and Public Policy: The Social and Ethical Problems of Fission Technology*, 2d ed. (Dordrecht: Reidel, 1983; original ed., 1980).

16. Friedrich Rapp and Paul T. Durbin, eds., *Technikphilosophie in der Diskussion* (Braunschweig and Wiesbaden: Vieweg, 1982); Paul T. Durbin and Friedrich Rapp, eds., *Philosophy and Technology*, Boston Studies in the Philosophy of Science, vol. 80 (Dordrecht: Reidel, 1983).

17. Fred I. Dretske, *Knowledge and the Flow of Information* (Cambridge, Mass.: MIT Press, 1981).

18. Friedrich Rapp, "Philosophy of Technology," in *Contemporary Philosophy: A New Survey*, vol. 2: *Philosophy of Science*, ed. G. Fløistad (The Hague: Nijhoff, 1982), pp. 361–412.

19. Carl Mitcham and Alois Huning, eds., *Philosophy and Technology II: Information Technology and Computers in Theory and Practice*, Boston Studies in the Philosophy of Science, vol. 90 (Dordrecht: Reidel, 1986).

20. Paul T. Durbin, ed., *Research in Philosophy & Technology*, vol. 7 (Greenwich, Conn.: JAI Press, 1984).

21. Joseph Agassi, *Technology: Philosophical and Social Aspects* (Dordrecht: Reidel, 1985); Albert Borgmann, *Technology and the Character of Contemporary Life: A Philosophical Inquiry* (Chicago: University of Chicago Press, 1985); Mario Bunge, *Treatise on Basic Philosophy*, vol. 7, pt. 2 (Dordrecht: Reidel, 1985); section 5 is entitled "Technology: From Engineering to Decision Theory" and contains Bunge's most elaborate formulation of an analytical philosophy of technology; Hans Lenk, *Zur Sozialphilosophie der Technik* (Frankfurt: Suhrhamp, 1982); Douglas Mac-Lean, ed., *Energy and the Future*, Maryland Studies in Public Policy (Totowa, N.J.: Rowman and Allanheld, 1983) (other volumes in this series cover such topics as air quality and risk assessment); Kristin Shrader-Frechette, *Science Policy, Ethics, and Economic Methodology: Some Problems of Technology Assessment and Environmental Impact Analysis* (Dordrecht: Reidel, 1984), and *Risk Analysis and Scientific Method: Methodological and Ethical Problems with Evaluating Societal Hazards* (Dordrecht: Reidel, 1985); Henryk Skolimowski, *Technology and Human Destiny* (Madras: University of Madras, 1983); Langdon Winner, *The Whale and the Reactor: A Search for Limits in an Age of High Technology* (Chicago: University of Chicago Press, 1986).

22. Staudenmaier, *Technology's Storytellers*, p. 201.

23. Ibid., p. 228, n. 57; see Hans-Georg Gadamer, *Truth and Method* (New York: Seabury, 1975; German original, 1960).

24. Mitcham, "Philosophy of Technology," and Carl Mitcham, "What Is the Philosophy of Technology?" *International Philosophical Quarterly* 25 (March 1985): 73–88.

25. Mitcham, "Philosophy of Technology," p. 288.

26. Mitcham, "What Is the Philosophy of Technology?" p. 73.

27. Mario Bunge is the most "serious"; see his *Treatise on Basic Philosophy*. See also three sample articles in *Research in Philosophy & Technology*, vol. 7, namely, Jose Felix Tobar Arbulu, "Plumbers, Technologists, and Scientists," pp. 5–17; Hans Lenk, "Toward a Pragmatic Social Philosophy of Technology and the Technological Intelligentsia," pp. 23–58; and Emmanuel Mesthene, "Technology as Evil: Fear or Lamentation?" pp. 59–74.

28. Mitcham, "Philosophy of Technology," pp. 343–44.

29. Thomas P. Hughes, *Networks of Power: Electrification in Western Society, 1880–1930* (Baltimore: Johns Hopkins University Press, 1983), p. 462.

30. Ibid.; and Jacques Ellul, *The Technological System* (New York: Continuum, 1980; French original, 1977). See also Ellul's earlier *The Technological Society*.

31. Langdon Winner, "*Technē* and *Politeia*: The Technical Constitution of Society," in Durbin and Rapp, *Philosophy and Technology*, pp. 97–111. See also Winner's two books, *Autonomous Technology* and *The Whale and the Reactor*.

32. David F. Noble, *America by Design: Science, Technology, and the Rise of Corporate Capitalism* (New York: Alfred A. Knopf, 1977).

33. Gendron, *Technology and the Human Condition*.

34. See n. 1 above.

35. Mitcham and Huning, *Philosophy and Technology II*.

36. The philosophical literature on AI is extensive. See Patrick H. Winston, *Artificial Intelligence* (Reading, Mass.: Addison-Wesley, 1977), and two surveys of philosophy of mind that include numerous references to the latest in AI literature: Daniel C. Dennett, "Current Issues in the Philosophy of Mind," and Jerome Shaffer, "Recent Work on the Mind-Body Problem (II)," both in *Recent Work in Philosophy*, eds. K. Lucey and T. Machan (Totowa, N.J.: Rowman and Allanheld, 1983), pp. 153–68 and 169–82.

37. The best summary of proposed reforms is in John Kenneth Galbraith, *Economics and the Public Purpose* (Boston: Houghton Mifflin, 1973).

38. Carol E. Hoffecker, *Corporate Capital: Wilmington in the Twentieth Century* (Philadelphia: Temple University Press, 1983).

39. John A. Munroe, *The University of Delaware: A History* (Newark: University of Delaware Press, 1986).

40. David Hounshell and John K. Smith, *Science and Corporate Strategy: DuPont R&D 1902–1980* (New York: Cambridge University Press, 1988).

41. The title of a collaborative volume spelling out the views of leading Pragmatists is *Creative Intelligence: Essays in the Pragmatic Attitude* (New York: Holt, 1917); John Dewey's lead essay was entitled, "The Need for a Recovery of Philosophy," pp. 3–69; three years later, he published *Reconstruction in Philosophy* (New York: Holt, 1920). George Herbert Mead also contributed a paper to *Creative Intelligence,* "Scientific Method and Individual Thinker," pp. 176–227; a more telling title is Mead's "Scientific Method and the Moral Sciences" (1923), but both essays espouse collaborative teamwork on social problems. Both are reprinted in *Selected Writings George Herbert Mead,* ed. A. Reck (Indianapolis, Ind.: Bobbs-Merrill, 1964).

Contextual Contrasts: Recent Trends in the History of Technology

MERRITT ROE SMITH
STEVEN C. REBER

The history of technology is a relatively new field of inquiry. As a scholarly specialty it can be traced, in this country, to the establishment of the Society for the History of Technology (SHOT) in 1958. Of course, the historical study of technology existed previously in the United States. Indeed, from the 1910s through the 1940s, numerous individuals became interested in various aspects of technological change and produced important works that now stand as classics. Among the most notable contributors were Louis C. Hunter, Lewis Mumford, Joseph W. Roe, Abbott P. Usher, and Lynn White, jr.[1] These scholars as well as others helped to pioneer the history of technology in America long before it became recognized as a bona fide field of specialization. Owing to their varied backgrounds in economics, engineering, history, and journalism, they tended to work in isolation from one another and lacked any shared notion of professional identity. The founding of SHOT gave continuity to the emerging field and provided its members with a special identity that had not existed before. Through the efforts of Melvin Kranzberg, the society's journal, *Technology and Culture,* became an important vehicle of scholarly communication and encouraged research in various new areas.

Impressive scholarly advances have been made in the history of technology during the past twenty-five years. Although attempts have been made throughout this period to report on the direction these advances have taken (and, at times, to suggest new directions), none has been as methodical as the analysis presented in John M. Staudenmaier's recent book, *Technology's Storytellers.* Using the articles published in *Technology and Culture* from 1959 through 1980 as his source material, he has studied the language,

An earlier version of this essay appeared in *Ceramics and Civilization,* vol. 3, *High-Technology Ceramics,* ed. W. David Kingery (Columbus, Ohio: The American Ceramic Society, 1986). The authors wish to acknowledge and thank the editor and the American Ceramic Society for permission to reprint it.

methodology, and predominant themes of scholarly writing in the history of technology. The result is a valuable historiographic survey that offers significant insights into the development of the field.[2]

Staudenmaier frames his discussion around the related concepts of design and ambience. By design he means the specific technical characteristics and constraints of a particular technology under study. Ambience, on the other hand, refers to the context (economic, social, political) in which the design emerges and develops. As Staudenmaier points out in an introductory chapter on methodological styles, histories of technology before the establishment of SHOT were dominated by an internalist methodology that addressed questions of design but largely ignored ambience. Singer, Holmyard, Hall, and Williams's encyclopedic *History of Technology* exemplifies this trend, with its disciplined attention to technical detail and to changes in design over time, but little or no discussion of the political or cultural contexts in which new technologies developed.[3] The founders of SHOT, concerned about the limits of internalism, made a concerted effort during the 1960s and 1970s to foster an alternative, contextual approach. The contextualist methodology stresses the importance of both design and ambience and attempts to understand their relationship.

Staudenmaier's analysis of *Technology and Culture*'s contents is divided into three chapters, each devoted to a major theme in the history of technology: emerging technology, the relationship between science and technology, and the interaction between technological development and culture. His discussion shows a fairly constant increase in both the number and sophistication of contextual studies in all three areas of discourse since the journal's inception. One effect of the mainstream acceptance of contextualism has been the emergence of a third methodology in the history of technology: externalism. In externalist studies a technology's ambience, not its design characteristics, forms the primary focus of discussion.

The growth and development of contextualism provides the main theme of Staudenmaier's book, and he links it explicitly to the philosophical debate over the ideology of progress.

> The myth of progress poses a particularly vexing problem for historians. Belief in an autonomous progress, beginning in Europe with the "scientific revolution" and proceeding with inevitable necessity in both scientific and technological domains ever since, implies a radical disjunction of method from context and, therefore, of technological design from human culture. (p. 164)

In other words, if technological progress is inevitable, then it is independent of context, and contextual issues can therefore have no explanatory power. Staudenmaier finds this "myth" objectionable and attributes the emergence of contextualism to the myth's growing unacceptability to most historians of technology.

Yet the idea of progress is not dead among historians of technology. Staudenmaier devotes his final chapter to the question of how far beyond "Whig history" (that is, the presentation of technological development as a linear success story) the discipline has progressed. His conclusions are mixed. Although impressed by the extent of contextual discourse among historians of technology, he finds the lack of research in some areas (technological failures, worker and feminist perspectives, non-Western technologies, and critiques of capitalism) to be an indication of methodological imbalance.

Staudenmaier's book is a perceptive, well-organized study that is of considerable use to both established scholars and newcomers to the history of technology. His insistence on the relationship between methodological positions and the ideology of progress, however, is bound to be controversial.

THE CONTEXTUALIST SPECTRUM

For the purposes of this essay, the main strength of Staudenmaier's book is his treatment of contextualism as an important (indeed, the most important) methodology available to historians of technology. His discussion thus provides an appropriate starting point for a review of some recent major works in the field. In the previous section, contextualism was defined as a recognition of the importance of cultural (typically social, political, and economic) factors that affect and are affected by technological development. This definition serves admirably to distinguish contextualism from internalist or externalist methodologies. It could be understood as implying that all scholars who describe themselves as contextualists bring the same approach to their studies. Such is not the case, however (as Staudenmaier himself notes). Indeed, a broad spectrum of approaches to contextualism is evident in the work of historians of technology. The ends of this spectrum correspond to two complementary yet quite distinct interpretations of technological change.

The older and more familiar of these interpretations is one that views technology as a form of expanding knowledge. Here scholarly discourse ranges from narrowly focused "nuts-and-bolts" studies to more complex treatments of innovation as well as the emergence of industrial systems. What distinguishes this approach is its primary emphasis on the evolving technical, managerial, and epistemological features of the subject. According to this perspective, cultural factors have importance chiefly as the background against which the main theme of technical development unfolds. Representative studies are David J. Jeremy's treatment of the transfer of technology in *Transatlantic Industrial Revolution,* David P. Billington's investigations of structural engineering in *The Tower and the Bridge,* Edward W. Constant's analysis of *The Origins of the Turbojet Revolution,* and Edwin T. Layton's research on the historical relations of science and technology as evidenced in "Mirror-Image Twins," and "Scientific Technology, 1845–1900."[4]

The second mode of interpretation concentrates not so much on the creative aspects of technological development, but rather on its social aspects: the reasons for developing a particular technology, the choices made among competing technologies, and the social implications of developing technologies. Here, the cultural context provides the dynamic force of the story, with technical factors in the background. The idea of technology as a "social force," implicit in this approach, has a rich tradition of scholarship behind it. Probably the best-known work in this area is Lynn White's seminal *Medieval Technology and Social Change,* although one can readily cite other studies such as Daniel R. Headrick's *Tools of Empire,* James J. Flink's *The Car Culture,* and various "retrospective technology assessments" that address the social and political implications of technological innovation.[5] Studies that view technology as a "social product" represent a more recent and controversial aspect of this second approach. Here the emphasis is on the attitudes, ideologies, and values that lie behind technology and, in effect, shape it. Works of this genre treat innovation in the workplace as well as more broadly defined analyses of technological change in local communities, geographic regions, and national cultures. Such an approach is reflected, for instance, in the writings of Edwin Layton, David F. Noble, Merritt Roe Smith, and Anthony F. C. Wallace.[6]

Four major recent works in the history of technology illustrate the variety of positions available on the contextualist spectrum: Thomas P. Hughes's *Networks of Power: Electrification in Western Society, 1880–1930;* David Hounshell's *From the American System to Mass Production, 1800–1932;* David F. Noble's *Forces of Production: A Social History of Industrial Automation;* and Ruth Schwartz Cowan's *More Work for Mother: The Ironies of Household Technology from the Open Hearth to the Microwave.* All four authors acknowledge the importance of studying technology within larger contextual settings. Each, however, occupies a distinctive position on the spectrum of contextualist methodology—hence the title of this essay, "contextual contrasts."[7]

Networks of Power is Thomas P. Hughes's comprehensive study of the development of electrical power systems in the United States and Europe in the period between 1880 and 1930. In the preface to the book, Hughes discusses his approach to the history of technology, placing himself squarely in the contextualist camp.

> I found that . . . with rare exceptions the impact of society, or culture, on the shape of technology had been virtually ignored. Dissatisfied with the internalist's approach, I turned to an exploration of a broad range of factors, events, institutions, men, and women involved in complex networks of power—technical, economic, political, and social. (p. x)

Hughes's goal is to explain the development of electrical power systems. For him, the defining characteristic of electrical power technology is its development as a system, a network of interconnected components, rather than as a collection of independent inventions. The systems concept is crucial to Hughes's argument, and the book is organized around a four-stage model of systems development. The stages are invention and development, technology transfer, systems growth, and systems momentum. The discussion of invention and development revolves around the work of Thomas Edison. According to Hughes's account, Edison's success in the field of electricity hinged on his systems approach to the problem. Rather than merely inventing the incandescent lamp, Edison realized from the outset that power generators, wiring networks, and light fixtures would all have to be developed before electric lighting would become a reality. This approach distinguished him from Joseph Swan, the British inventor of the incandescent lamp, who was concerned only with a single component.

The problems of technology transfer, the second stage of Hughes's model, are illustrated by the attempts to transplant American electrical power technology to England and Germany in the 1880s. Here political and economic forces come into play that influence the shaping of systems. The political realities in England, particularly the regulatory impulses of the Gladstone government, resulted in legislation that originally priced electricity out of the lighting market. In Germany, after initial conflicts between growing German nationalism and the importation of American technology were resolved, the transfer of technology was successfully effected. Hughes's discussion of the third stage, systems growth, likewise relies on a comparative study of power systems, specifically those in Berlin, Chicago, and London. In Berlin, political and business forces compromised, and systems growth depended on their continued cooperation. In Chicago, the dominance of Samuel Insull and other manager-entrepreneurs over local political authority resulted in an absence of political limitations on systems growth, and a relative increase in the importance of technological and economic limitations. In London, the reverse was true: political forces placed much sharper restrictions on systems growth than technological or economic conditions.

According to Hughes, as a technological system grows, it develops an internal dynamic that makes it difficult for external factors to change the direction of its evolution. This is the fourth stage, that of technological momentum. In the field of electric power, this momentum was due to large amounts of capital investment, the growth of electrical manufacturing firms, the development of electrical engineering education, and the activities of professional engineering societies. (Only a cataclysm like World War I could change the direction of independent American power companies towards cooperation. But with the end of the war, the industry returned to its previous direction of development.) At this stage the primary obstacles were financial: for example, the development of regional power systems required large-scale capital formation. This critical problem was solved by the emergence of large holding companies, pioneered by financier-entrepeneurs like S. Z. Mitchell.

In *Networks of Power* Hughes presents a rich and complex study of the development of electrical light and power technology. The scope of the study is impressive, and his findings are convincing. The nature and presentation of these findings offer some clues to Hughes's underlying methodology.

The theme of the book is the emergence of a dynamic technological system against a background of economic, political, and geographical constraints. Hughes's four-stage model of system development is one of fairly steady linear progress punctuated by temporary obstacles ("reverse salients"), both technical and cultural in nature. The narrative focus is on the leading entrepreneurial figures of the electrical power industry, individuals like Edison, Insull, and Mitchell. The analysis is thus from the top down.

Although Hughes notes that "of the great construction projects of the last century . . . none has been more influential in its social effects . . . than the electrical power system" (p. 1), these effects, and their implications for consumers of electricity or workers in the electrical industry, are not a part of this study. No reference is made, for example, to linemen or other lower-level employees who played essential roles in constructing and maintaining large electrical power systems. Hughes makes it clear from the outset that the problem under analysis is the development of electrical power technology itself, not labor-management relations or the larger social implications of technological change. The result nonetheless is a fascinating and original work located squarely at the end of the contextual spectrum concerned with "technology as expanding knowledge."

The second book, David Hounshell's *From the American System to Mass Production,* treats the diffusion and assimilation of new technologies among large-scale manufacturers. Specifically, it focuses on how innovations that originated in the antebellum American firearms industry spread to other technically related industries during the nineteenth century and, with further elaborations and additions, eventually culminated in Henry Ford's system of mass production in 1913.

Hounshell's methodological approach has much in common with that of Hughes. Both focus on technical creativity, innovation, and their relationship to the evolution of large industrial systems, and both explicitly reject the idea of unilinear technological progress. Hounshell offers a "diffusion" model, however, to account for the spread of manufacturing innovations among industries involved in activities as seemingly diverse as sewing machine manufacture and automobile mass production.

In detailing how various innovations filtered from firm to firm and industry to industry, Hounshell breaks new ground. Historians have long realized that a genealogical connection existed between arms-making and other metal-working industries in the United States, but they knew relatively little about how it actually took place. Hounshell makes these connections clear for clock, sewing machine, furniture, reaper, bicycle, and automobile manufacturing, and he does so in exquisite detail. What is more, he reveals how inherently complicated the process of change and adaptation really was. Instead of relating heroic feats of genius, Hounshell chronicles the aggravating, frustrating, and mundane aspects of innovation. In his superb second chapter on sewing machine production, for example, he argues that the decisions of some manufacturers about whether to adopt practices from the armories often depended on personalities and work traditions within individual firms. Hence the Wheeler and Wilson sewing machine company of Bridgeport, Connecticut, rapidly adopted armory practices primarily because its superintendent (W. H. Perry) and several key mechanics originally came from the firearms industry. Indeed, Hounshell notes, "the small Wheeler and Wilson factory seemed a microcosm of Colt's armory" in nearby Hartford (p. 70).

The Singer Manufacturing Company, by far the largest and most famous sewing machine maker, presented a marked contrast to Wheeler and Wilson. Because the firm's owners and overseers were unacquainted with "armory practice," over thirty years elapsed before interchangeable techniques replaced Singer's original labor-intensive "European method" of manufacturing. The same was basically true of one of the leading agricultural machinery manufacturers of the era, the McCormick reaper works in Chicago. Not until 1880, thirty-two years after its establishment, did the company turn to production methods used in small-arms plants, and then only after the firm's founder, Cyrus McCormick, fired his craft-trained brother as shop superintendent and replaced him with a former Colt employee. Whether the subject is sewing machines, agricultural implements, or automobiles, Hounshell is clearly correct when he observes that "the process of diffusion was neither as smooth nor as simple . . . as others would have it" (p. 5).

But that is only part of the story. Even "best practice" firms such as Brown and Sharpe (manufacturers of the Willcox & Gibbs sewing machine), the Pope Manufacturing Company (maker of the Columbia bicycle), and the Ford Motor Company continually experienced difficulties in tooling up for production and maintaining high standards once production was underway. In short, the adoption of new techniques in other industries proceeded, like they had at the armories themselves, by fits and starts. On more than one occasion, the Ford Company was beset by design delays and production bottlenecks in its touted mass-production system. Indeed, Hounshell describes the changeover from the Model T to the Model A Ford as a chaotic "comedy of errors" (p. 191). The impression one comes away with is how bumpy and twisted the path of technological progress often is.

As with Hughes's *Networks of Power*, Hounshell emphasizes technical development, with cultural context providing the background. Hounshell is not content to rest firmly in the camp that sees "technology as expanding knowledge," however, and he makes explicit note of the social implications of his subject. He concludes the book with a chapter about the "Ethos of Mass Production" in which he discusses the positions held by various proponents of the new factory system. Indeed, his treatment of such social commentators as businessman Edward Filene, actor Charles Chaplin, and artist Diego Rivera represents one of the most absorbing parts of the book. Yet, as interesting as these sketches are, the reader leaves the chapter with a rather unfulfilled feeling—perhaps because Hounshell neglects to address the day-to-day social relations of mass production, particularly the worker's response to it. Although he excels at describing the means by which new techniques moved from firm to firm and industry to industry, his analysis stops short of treating what actually happened on the shop floor. Throughout the book he acknowledges that the introduction of mass-production techniques "brought serious labor problems" (pp. 11, 182, 307), yet at no point does he pause to analyze and assess this critical aspect of technological change. As arresting as the art of Chaplin and Rivera are, their work is too artificial and neat to capture the messy human relations of production within factory settings. For all his attention to the social processes of innovation and diffusion, Hounshell's approach, like Hughes's, remains essentially an entrepreneurial one.

In David Noble's book, *Forces of Production,* the contextual emphasis shifts to the social factors that lie behind technological innovation. Specifically, he attacks two widely held and closely associated beliefs about modern technological development. One is the popular notion, well known among scholars as technological determinism, "that machines make history rather than people" (p. 324). The other is the "Darwinian view" of technological development that "assumes that the flow of creative invention passes progressively through three successive filters"—objective science, economic rationality, and the self-regulating mechanism of the market—thus assuring that only the fittest technologies survive. Both positions, Noble argues, tend to objectify technology and set it apart from human affairs with the result that the social intentions, interests, and choices that lie behind technological innovation are either overlooked or dismissed as irrelevant. Against these views Noble arrays considerable evidence showing that those who decide what new technologies will be adopted "are not autonomous agents of some disembodied progress" but rather members of society who "are moved, like everyone else, by a myriad of motivations" that reflect cultural values, political ideologies, economic interests, intellectual enthusiasms, professional identities, and the like (p. 43). In effect, Noble contends that "the process of technological development is essentially social" (p. xiii). To illustrate how social choices influence technological change, he turns to "the guts of modern industry," the design and introduction of automatically controlled machine tools.

Much of the story centers on the efforts of John T. Parsons, president of one of the country's primary manufacturers of helicopter rotor blades, to develop numerically controlled machinery for cutting integrally stiffened wing sections of high-speed military aircraft. Having received a U.S. Air Force contract in June 1949 to design and build an "automatic contour cutting machine" with three-axis capability, Parsons subcontracted with the Massachusetts Institute of Technology's Servomechanisms Laboratory to develop electronic controls for the unit. Yet, no sooner had MIT entered the project than Parsons began to be eased out. Noble argues that the Servo Lab's director and other "high-powered" MIT engineers were looking for "new worlds to conquer" and before long transformed Parsons's original idea into a much more ambitious program for research, development, and private gain (p. 116). With encouragement and funding from the Air Force, the MIT group ended up developing an extremely sophisticated continuous-path, five-axis milling machine programmed by an extremely complex and expensive automatically programmed tools (APT) system. Altogether the Air Force spent more than sixty million dollars on the development project at MIT between 1949 and 1959 and at least another hundred million on buying and installing numerically controlled machinery in the factories of private aerospace contractors. With such a heavy government investment in the new technology, machine tool builders (who were initially skeptical) eventually joined "the rush" to numerical control (NC) and made it the industry standard.

At the very time NC was taking shape at MIT, other aspiring inventors were trying to introduce similar types of automatically controlled machinery. One of the most promising efforts took place at General Electric in Schenectady, New York, where between 1946 and 1953 a team headed by Lowell Holmes and Lawrence Peaslee perfected a computer method for guiding machine tools called "record-playback" (R/P). But, according to Noble, R/P and other innovative techniques were eventually abandoned because they had the "unfavorable feature" of being programmed and controlled by workers on the shopfloor (p. 151). Unlike NC, which allowed for programming in the central office by management, R/P was too accessible. The approach proved unattractive to military-industrial decisionmakers because it gave machinists too much freedom to determine the pace of their work and did not "contribute toward close production control by management" (p. 152).

The desire to increase managerial control appears to be the primary reason for the rejection of R/P, but it was not the only one. Noble points to a constellation of factors that favored NC, not least of which was a compulsive fascination, shared equally by Air Force officers and MIT engineers, with automaticity and remote control. This enthusiasm for automation, coupled with MIT's political influence, a cold war-spawned fear of falling behind the Russians, a basic distrust of labor, and a desire to extend standardized manufacturing methods, provided the "overriding impulse" behind the push for numerical control (p. 334). "People, not destiny, chose this technology," Noble concludes, "and their combined power . . . overwhelmed the alternatives" (p. 146). Consequently when R/P entrepreneurs approached the Air Force for developmental funding, they were turned down. "What money there was went into numerical control" (pp. 177, 192).

As with the works of Hughes and Hounshell, an important connection exists between Noble's conclusions and his underlying methodology. In this respect, it is significant that Noble considers himself a social historian rather than, strictly speaking, a historian of technology. He makes it clear from the outset of his book that his primary interest is in an examination of the related myths of technological determinism and technological "evolution," myths that he believes have important social and political ramifications. Thus, he is more interested in the social influences affecting technological development, and the social implications of that development, than in the "nuts and bolts" of any particular technology. The contrast with Hughes and Hounshell is striking.

This social emphasis has both positive and negative results. It leads Noble to consider alternative technologies in the automatic control field and to pay close attention to their implications for the workers who must use them, topics that (as Staudenmaier has pointed out) rarely receive close analysis by historians. On the other hand, his technical discussion of automatic control, while informative, does not approach the mastery of detail attained by Hughes in the area of electrical power.

Forces of Production is a rich and absorbing book, the product of a prodigious research effort. Indeed, its subject matter is so fresh, so original, that there is yet no comparable body of scholarship against which it can be measured. Consequently, it remains to be seen just how well Noble's thesis about the origins of numerical control will fare in the world of scholarship.

Ruth Schwartz Cowan's book looks at a different labor force: those who work in the home. *More Work for Mother* presents an interesting alternative to the traditional view of the home as "the last gasp of feudalism," a preindustrial haven from the capitalist world (p. 5). Cowan argues that technological development has changed the home as drastically as it has changed industry. Her study deals with the United States, and divides the history of American homes into three periods: preindustrial (to about 1850), early industrial (1850–1900), and twentieth century.

The key to Cowan's analysis lies in her concept of the work process. To treat housework as a collection of simple, separable tasks (cooking dinner, cleaning rugs, attending sick children) is misleading, she contends. Cooking, for example, is a procedure that involves much more than just stirring the pot. Food storage, the provision of an energy source, the maintenance and cleaning of cooking tools, and the disposal of waste products are all integral parts of the cooking "work process." To demonstrate the changes wrought by industrialization, Cowan describes in detail typical work processes in each of the three periods of her study. In the preindustrial era, housework in the average home was shared among wife, husband, and children or servants in a recognizable division of labor. During this largely agrarian period, men provided basic foodstuffs in the form of grains, butchered meat, or game; in addition they cut wood for fuel, prepared textile fibers for spinning, and fashioned basic household tools from available materials. Women grew vegetables, prepared meals, made clothing, and bore children. Servants or older children performed cleaning chores, carried water from wells or streams, and minded the small children.

With the rise of industrialism, this scene underwent significant changes. The division of industrial labor meant that the husband often left the home and small farm to earn cash to buy food and household implements. The implements themselves underwent striking technological changes. The introduction of coal stoves, far more efficient than wood fires, meant that less fuel had to be provided. The availablity of manufactured cloth eliminated the need for weaving and spinning. Piped household water put an end to water carrying. Cowan summarizes the effect of industrialization on the home as "more chores for women, fewer for men" (p. 63); almost all of these technological changes reduced the household labor previously performed by husbands, servants, or children. The work of the wife, on the other hand, was augmented by the changes: coal stoves, unlike fireplaces, required considerable cleaning; clothes made from manufactured cloth, unlike those made from leather or wool, required laundering. In addition, servants began to leave homes during this period for higher factory wages; their work was assumed by the housewife.

The pace of change quickened during the twentieth century. Electricity, improved health care, and the automobile were among the most visible factors affecting housekeeping in the first half of the twentieth century. Cowan's discussion of this period employs a "systems" concept of emerging technology not unlike that of Hughes. Rather than discussing individual innovations in household technology in isolation, she defines eight "interlocking" systems: food, clothing, health care, transportation, water, gas, electricity, and petroleum products. Several of these systems (food, clothing, health care) follow the conventional model, in which industrialization displaces home production with home consumption. The advent of the automobile, however, offers a case in which the effect was reversed. Before 1900, households were consumers of transportation; many purchased goods were delivered or peddled door-to-door. If family members needed transportation, they often relied on privately hired vehicles or public transit. With the introduction of mass-produced automobiles, all this changed. Shopping, the wife's job, increasingly involved a drive to the store. Personal transportation for family members also made use of the automobile, and the job of chauffeur was soon added to the housewife's daily chores.

In the cases of the remaining systems, the effects were ambiguous; at first glance, technological development in these water and energy utilities led simply to an increase in available "labor-saving" devices. However, the results often involved less drudgery but more work: more laundering, because homes had their own washing machines readily available; more cleaning, thanks to the appearance of the bathroom; more involved meal preparation, because a wider variety of foods was available. Cowan also notes a change in the popular image of the mother, a change spawned partly by aggressive advertising by the manufacturers of the new labor-saving devices. This new image emphasized the mother's "duties" (and often played on feelings of guilt): keeping the home and clothing spotlessly clean, the children glowing and healthy, and the meals nutritious and appetizing. The combination of technological development and social forces led inexorably to "more work for mother."

Cowan is not claiming that the new household technology, by eliminating most male tasks, left men free to join the industrial work force; the direction of causality lies clearly in the opposite direction. She is, however, addressing a similar claim that technological and social developments in these areas reached a point where women were free to leave the home. As the preceding summary indicates, Cowan disagrees strongly with this notion. She strengthens her claim by describing an alternative path of development that would have truly enabled women to put housework aside. In her chapter on alternative social and technical approaches to housework, she provides an interesting outline of some innovations in housework that failed to achieve widespread popular support: the commercial laundry, the cooked-food delivery service, and the apartment hotel. These cooperative services, and others like them, could have almost completely eliminated household chores, yet they were largely rejected. The reasons, Cowan explains, were not economic; the alternatives cost no more. On the contrary, they were social: "When given choices . . . most Americans act so as to preserve family life and family autonomy" (p. 150).

Cowan shares with David Noble an interest in technology as a social force, and an emphasis on social, rather than technical, dynamics. The actors in her account, far from being heroic figures, are household workers, average members of the population whose stories are gleaned from diary entries and family budget records. Although her technical discussions are sophisticated, they are not detailed; such detail is reserved for descriptions of household life during the different periods of American history. Unlike Noble, Cowan does not analyze the social production of technology; for the purposes of her study, technological development is largely an independent variable. Despite this emphasis, her brief exposition on alternative household technologies seems to support her belief that technological development is at least partially socially determined.

In *More Work for Mother* Cowan has taken on a complex problem and solved it in an original and interesting way. While answering many questions about the relationship between industrialization and housework, she has raised many more, showing the importance of a technological perspective in what is, as yet, a developing area of historical research. Like Noble's work, this book will certainly be a powerful stimulus to further research.

CONCLUSION

In the final chapter of *Technology's Storytellers,* John Staudenmaier offers the following observation about the emerging contextualist methodology.

> Genuine contextualism is rooted in the proposition that technical designs cannot be meaningfully interpreted in abstraction from their human context. The human fabric is not simply an envelope around a culturally neutral artifact. The values and world views, the intelligence and stupidity, the biases and vested interests of those who design, accept, and maintain a technology are embedded in the technology itself. . . . Contextual history of technology affirms as a central insight that the specific designs chosen by individuals and institutions necessarily embody specific values. (pp. 165–66)

In this essay we have reviewed four recent works that belong in the general category of contextual histories of technology. Each has been enriched by attention to the cultural context surrounding technological development. We have distinguished between two basic perspectives among contextualists: the first, which views technology as expanding knowledge, is quite evident in the work of Hughes and Hounshell. Their books provide splendid examples of how this older, more conventional approach to the history of technology can be enriched by treating the subject from a contextual perspective. The more recent approach, which views technology as both a social product and a social force, is employed in the works of Noble and Cowan. Their works provide provocative examples of how social forces enframe, and may actually define, technological change.

In identifying the predominant approaches of these authors, the intent is not to favor one approach over the other, or to imply that one completely precludes the other. On the contrary, elements of both approaches are evident at various points in all four works. Nevertheless, the identification of an author's basic methodology is an important part of the critical process. Staudenmaier's words apply to historical perspectives as well as technical designs. They also imply values and world views, biases and vested interests. This is the nature of historical research, or indeed of any other interpretive endeavor.

The approach of Thomas Hughes and David Hounshell reflects an attachment to the technically oriented foundations of the history of technology.

Interpretations of design and innovation receive rigorous critical attention in their work, whereas interpretations of social, political, or economic institutions receive less emphasis. Inventors, managers, and business people are the important actors in technological development, according to this view, and their motivations are matters of assumption rather than the subject of inquiry. The strength of this approach, then, is the breadth and complexity it achieves by its concentrated attention on entrepreneurial behavior and technical detail.

David Noble and Ruth Cowan reside at the other end of the contextual spectrum. They are addressing cultural issues that they believe cannot be properly studied if technological issues are not addressed. Both authors place primary emphasis on workers: machine-tool operators in Noble's study, houseworkers in Cowan's.

It is clear that, of the two perspectives we have discussed, this latter one is the most recent and the most controversial. It is controversial because it treats inherently complex and sensitive issues such as the relationships of technology to class interest, economic power, and social change. If the "expanding knowledge" approach can be said to exhibit a degree of political and social conservatism, the "social product/social force" approach may easily be suspected of radicalism. Both Noble and Cowan present conclusions that depart significantly from mainstream analysis in their respective areas of study.

Several leading historians of technology have expressed misgivings about the new perspective. Probably the most candid statement of this position comes from Brooke Hindle, the esteemed senior historian of technology and former director of the Smithsonian's National Museum of American History. In a recently published review essay entitled "The Exhilaration of Early American Technology: A New Look," Hindle voices concern about what he calls "the dark side" of the history of technology and the threat it poses to the development of the field.[8] Although he does not reveal who these "darksiders" are, he apparently identifies them with the new social history and the breakup of the "consensus school" of interpretation that dominated American historical studies from the end of World War II to the mid-1960s. His greatest concern, however, is that darkside historians who seem to dwell on "failures, defeats, and negatives" have had a disruptive effect on attempts to achieve synthesis in either the history of technology or American history more broadly. Indeed, he remarks that the general framework of American history "has been so damaged by the challenge of the dark side . . . that an acceptable synthesis has become hard even to imagine."[9] The operative word here, of course, is "acceptable." Not everyone in the history of technology field views the social interpretation of technology in such a negative light, nor do they believe that it has inhibited our ability to achieve acceptable syntheses either in American history or the history of technology. Our view is that it has helped to enrich our understanding of technology in human affairs and to refine our perceptions of social progress. What is more, the social approach is helping to break down intellectual barriers that have long separated the history of science and technology from general history. Until recently few historians paid serious attention to the history of technology because it seemed "too technical" or, worse, too remote from their primary interests in politics and society. The dismal chapters on industrialization that proliferate in most American history textbooks readily testify to the degree of this neglect. Yet, despite this unsatisfactory state of affairs, there is reason for optimism when one reads books such as those reviewed here. From them we begin to realize that the big questions in the history of American technology are not all that different from the big questions in American history. Ideally both should treat the diverse and competing traditions that have shaped the country's history. To do so, both must grapple with how society works by identifying diachronic patterns of thought and action and analyzing the social processes that underlie them.

The social interpretation of technology has had a net beneficial effect on the field because it addresses these questions in a more forthright manner than the older internalist, consensus tradition in the history of technology. Precisely because it does this in a way that is more meaningful to historians outside the field, the social perspective holds the greatest promise for achieving a common dialogue within the larger profession and ultimately, for placing technological development in the mainstream of history.

This said, there are equally good reasons for maintaining the more traditional emphasis on technology as expanding knowledge; it remains the center scaffolding of the history of technology, and continues to motivate the work of most scholars in the field. Our hope is that historians of that persuasion will follow the example of Hounshell and Hughes and pay greater attention to the social determinants of technological change, just as we hope social historians will follow the example of Noble and Cowan in making use of technological studies. This integration is needed, and the authors considered here provide stimulating examples of how it can be achieved.

NOTES

1. See Louis C. Hunter, *Steamboats on Western Rivers* (Cambridge, Mass.: Harvard University Press, 1949); Lewis Mumford, *Technics and Civilization* (New York: Harcourt, Brace & World, 1934); Joseph W. Roe, *English and American Tool Builders* (New Haven, Conn.: Yale University Press, 1916); Abbott P. Usher, *A History of Mechanical Inventions* (Cambridge, Mass.: Harvard University Press, 1929); and Lynn White, jr., *Medieval Technology and Social Change* (New York: Oxford University Press, 1962). Other significant early works include Greville and Dorothy Bathe, *Oliver Evans* (Philadelphia: The Historical Society of Pennsylvania, 1935); Roger Burlingame, *The March of the Iron Men* (New York: Charles Scribner's Sons, 1938) and *Engines of Democracy* (New York: Charles Scribner's Sons, 1940); Howard I. Chapelle, *The History of the American Sailing Navy* (New York: Norton, 1949); Waldemar Kaempfert, ed., *A Popular History of American Invention,* 2 vols. (New

York: Charles Scribner's Sons, 1924); John A. Kouwenhoven, *Made in America: The Arts and Modern Civilization* (Garden City, N.J.: Doubleday, 1948); Leo Rogin, *The Introduction of Farm Machinery in Its Relation to the Productivity of Labor in the United States during the Nineteenth Century* (Berkeley: University of California Press, 1931); David B. Steinman, *The Builders of the Bridge* (New York: Harcourt, Brace, & Co., 1945); Dirk J. Struik, *Yankee Science in the Making* (Boston: Little, Brown and Company, 1948).

2. John M. Staudenmaier, S. J., *Technology's Storytellers: Reweaving the Human Fabric* (Cambridge, Mass.: MIT Press, 1985). Other historiographic discussions may be found in Angus Buchanan, "Technology and History," *Social Studies of Science* 5 (1975): 489–99; George Bugliarello and Dean B. Doner, eds., *The History and Philosophy of Technology* (Urbana: University of Illinois Press, 1979); Donald S. L. Cardwell, "The Academic Study of the History of Technology," *History of Science* 7 (1968): 112–24; George H. Daniels, "The Big Questions in the History of Technology," *Technology and Culture* 11 (January 1970): 1–21; Eugene S. Ferguson, "Towards a Discipline of the History of Technology," *Technology and Culture* 15 (January 1974): 13–30; Brooke Hindle, "The Exhilaration of Early American Technology," in *Technology in Early America* (Chapel Hill: University of North Carolina Press, 1966): 3–28; Brooke Hindle, "A Retrospective View of Science, Technology, and Material Culture in Early American History," *William and Mary Quarterly* 41 (July 1984): 422–35; David A. Hounshell, "On the Discipline of the History of Technology," *Journal of American History* 67 (March 1981): 854–65; David A. Hounshell, ed., "The History of American Technology: Exhilaration or Discontent?" *Hagley Papers* (Greenville, Del.: Hagley Museum and Library, 1984); Thomas P. Hughes, "Emerging Themes in the History of the Technology," *Technology and Culture* 20 (October 1979): 697–711; Edwin T. Layton, Jr., "Technology as Knowledge," *Technology and Culture* 15 (January 1974): 31–41; Carroll W. Pursell, Jr., "History of Technology," in *A Guide to the Culture of Science, Technology, and Medicine*, ed. Paul T. Durbin (New York: Free Press, 1980, 1984); Carroll W. Pursell, Jr., "The History of Technology and the Study of Material Culture," *American Quarterly* 35 (1983): 304–15; Reinhard Rürup, "Historians and Modern Technology," *Technology and Culture* 15 (April 1974): 161–93; Darwin H. Stapleton and David A. Hounshell, "The Discipline of the History of American Technology: An Exchange," *Journal of American History* 68 (March 1982): 897–902; Lynn White, jr., "The Discipline of the History of Technology," *Journal of Engineering Education* 54 (1964): 351.

3. Charles Singer, E. J. Holmyard, A. R. Hall, and Trevor I. Williams, eds., *History of Technology*, 5 vols. (London: Oxford University Press, 1954–58).

4. David J. Jeremy, *Transatlantic Industrial Revolution: The Diffusion of Textile Technologies between Britain and America, 1790–1830* (Cambridge, Mass.: MIT Press, 1981); David P. Billington, *The Tower and the Bridge* (New York: Basic Books, 1983); Edward W. Constant, *The Origins of the Turbojet Revolution* (Baltimore: Johns Hopkins University Press, 1980); Edwin T. Layton, Jr., "Mirror-Image Twins: Communities of Science and Technology in Nineteenth Century America," *Technology and Culture* 12 (October 1971): 562–80; Edwin T. Layton, Jr., "Scientific Technology, 1845–1900: The Hydraulic Turbine and the Origins of American Industrial Research," *Technology and Culture* 20 (January 1979): 64–89.

5. White, *Medieval Technology and Social Change;* Daniel R. Headrick, *The Tools of Empire: Technology and European Imperialism in the Nineteenth Century* (New York: Oxford University Press, 1981); James J. Flink, *The Car Culture* (Cambridge, Mass.: MIT Press, 1975). On retrospective technology assessment, see, e.g., Bruce Mazlish, ed., *The Railroad and the Space Program* (Cambridge, Mass.: MIT Press, 1965); Joel A. Tarr, ed., *Retrospective Technology Assessment* (San Francisco: San

Francisco Press, 1976); Joel A. Tarr and Francis C. McMichael, *Retrospective Assessment of Wastewater Technology in the United States: 1800–1872* (Springfield, Va.: National Technical Information Service, 1977); Joel A. Tarr, ed., "The City and Technology," *Journal of Urban History* 5 (May 1979): 275–406; Ithiel de Sola Pool, *The Social Impact of the Telephone* (Cambridge, Mass.: MIT Press, 1977); Vary T. Coates and Bernard Finn, *A Retrospective Technology Assessment: Submarine Telegraphy—The Transatlantic Cable of 1866* (San Francisco: San Francisco Press, 1979); and Martin V. Melosi, ed., *Pollution and Reform in American Cities, 1870–1930* (Austin: University of Texas Press, 1980). For historiographic assessments of retrospective technology assessment, see Howard P. Segal, "Assessing Retrospective Technology Assessment," *Technology in Society* 4 (1982): 231–46; and Stephen Cutcliffe, "Retrospective Technology Assessment," *STS Newsletter* (Lehigh University) 18 (June 1980): 7–12.

6. Edwin T. Layton, Jr., *The Revolt of the Engineers: Social Responsibility and the American Engineering Profession* (Cleveland: Case Western Reserve University Press, 1971; Baltimore: Johns Hopkins University Press, 1986); David F. Noble, *America By Design: Science, Technology, and the Rise of Corporate Capitalism* (New York: Alfred A. Knopf, 1977); Merritt Roe Smith, *Harpers Ferry Armory and the New Technology* (Ithaca, N.Y.: Cornell University Press, 1977); Anthony F. C. Wallace, *The Social Context of Technology* (Princeton: Princeton University Press, 1982); and Anthony F. C. Wallace, *Rockdale: The Growth of an American Village in the Early Industrial Revolution* (New York: Alfred A. Knopf, 1978).

7. Thomas P. Hughes, *Networks of Power: Electrification in Western Society, 1880–1930* (Baltimore: Johns Hopkins University Press, 1983); David Hounshell, *From the American System to Mass Production, 1800–1932* (Baltimore: Johns Hopkins University Press, 1984); David F. Noble, *Forces of Production: A Social History of Industrial Automation* (New York: Alfred A. Knopf, 1984); Ruth Schwartz Cowan, *More Work for Mother: The Ironies of Household Technology from the Open Hearth to the Microwave* (New York: Basic Books, 1983).

8. In Hounshell, "The History of American Technology," pp. 7–17. Also see Hindle's essay in this volume.

9. Hounshell, "The History of American Technology," pp. 8, 12.

The Politics of Successful Technologies

JOHN M. STAUDENMAIER, S.J.

> America's great evasion lies in the manipulation of nature to
> avoid a confrontation with the human condition and with the
> challenge of building a true community.
>
> William Appleman Williams

William Appleman Williams's provocative statement confronts citizens of
the United States with the link between technological style and national
character.[1] Has the American technical dream of conquering the wilderness
encouraged escape from the essentially public and political labor of negotiating the terms for human community? Historians at least as far back as
Frederick Jackson Turner have wrestled with the question.[2] Recently we find
a related issue addressed in sociological discourse. Robert Bellah and his
coauthors in *Habits of the Heart* speak of American individualism grown
"cancerous" as the balancing traditions of republicanism and Christianity
recede in the national consciousness.[3] Their extended interviews with more
than two-hundred middle-class men and women reveal a poverty of language
when Americans try to explain their behavior and motivation in terms that
include a vision of the common good.

In like fashion, communications scholar James W. Carey sees several
clashing themes—technology versus community and power versus tran-
quility—as the essence of a peculiarly American dualism.

> The American character since early in its history has been pulled in two
> directions and has been unable to commit itself to either. The first
> direction is toward *the dream of the American sublime, to a virgin land
> and a life of peace, serenity and community*. The second direction is *the
> Faustian and rapacious, the desire for power, wealth, productivity and
> universal knowledge, the urge to dominate nature and remake the world*.
> In many ways the American tragedy is that we want both these things
> and never seem to respect the contradiction between them. (Emphasis
> added.)

Carey attributes the staying power of the contradiction to a gradual cognitive shift from an idealization of Nature to an even greater idealization of "the technological sublime."

> The dream of the American sublime has never been used to block the dream of American power primarily because the rhetoric of the technological sublime has collapsed the distinctions between the two directions. As a result, America has allowed technology and the urge for power unrestricted room for expansion.[4]

For Carey, as for Williams, Americans have substituted the sweet promise of technological progress for the inherently uncertain process of negotiating community.[5] Their hypothesis helps explain Bellah's observation that contemporary Americans find it hard to move from private to public discourse. Together, they call attention to a social problem of substantial magnitude. Does the common good take care of itself while individual Americans pursue their own ends? If not, where might we look for a language that helps us to work at articulating a public vision of national identity?

How might historians of technology contribute to such a linguistic search? By raising this last question, I imply that the Society for the History of Technology (SHOT) has reached a level of institutional maturity wherein it is no longer primarily, or at least exclusively, concerned with its internal life. By their increasing willingness to engage thorny political and ethical questions as they delineate research topics, SHOT members have defined themselves as actors within the public domain.[6] In what follows, then, I will argue that historians of technology have an important role in late twentieth-century civic discourse and that creation of language, always the historian's contribution to society, is badly needed today precisely in the area of technology-society dynamics.

We might begin by noting that the ideology of progress legitimates America's substitution of technology for political negotiation. As I argued in *Technology's Storytellers,* autonomous technological determinism rests on the radically ahistorical premise that technical designs operate according to their own inevitable laws, that the context of origins has little to tell us about the design decisions that bring new technologies into being.[7] If such progress sustains the American dream, then we need look no further to reassure ourselves about the health of our technocratic social venture.

In sharp contrast, the contextual approach to the history of technology, pioneered by SHOT in the last quarter-century, challenges progress ideology by situating technical design decisions within the public fabric of societal decision making. Historians of technology have begun looking to the values, biases, motives, and world view of the designers when asking why a given technology turned out as it did. From the contextualist perspective, technologies are political and cultural as well as technical artifacts. Historical interpretation of successful technologies engages in social history at a foundational level, especially when studying a society that has adopted such an extraordinary array of complex systems as the governing structures of its culture.[8]

In *Storytellers* I argued that an overall model for the contextual study of successful technologies includes three stages in the technological life cycle—namely, the design stage, the momentum stage, and the senility stage. All three reveal the contextual character of the technical design as it moves from the inherent fragility of its time of origin through a period of successful momentum and into obsolescence. I argued further that successful technologies come in clusters, each embodying the same set of values, the entire cluster reflecting the technological style of its culture of origin. Finally, I suggested that when historians interpret technological style they must pay attention to three groups as they relate to the technology in question: the design, maintenance, and impact constituencies.[9]

In the years since proposing the three-constituency model, I have wrestled with its limitations in two areas. On the one hand, its focus on a single successful technology tends to create the simplistic impression that new technologies can be understood without reference to the tradition of the technological style that preceded it or to technological styles that did not achieve societal acceptance. On the other hand, *Storytellers'* description of the design, maintenance, and impact constituencies overlooked a crucial dimension of their technological relationship—the fact that members of each technological constituency tend to mask their connection with a successful technology.

In this essay I will explore both aspects of the model using the American automobile, whose popularity matches its historical familiarity, as an exemplary case study. After tracing the broad outlines of automotive history from 1900 to the present, I will return to the nineteenth century, situating automotive history within a larger technical and social context—America's adoption of the ideal of standardization and its correlative turning from a negotiation style for solving technological problems. Finally, I will return to the design, maintenance, and impact constituencies in an attempt to explain why their relationships with successful technology prove so elusive. This essay, then, blends social analysis and historiography. It rests on the premise that respect for the complexity of technological style is important both for historians of technology and for Americans generally. The ideology of autonomous progress, with its seductive simplicity, may well prove to be a major reason why Americans find it so hard to interpret their experience in communal language. By their pursuit of a contextual method, historians of technology can play a significant role in creating a healthier body of social discourse.

PART I: STAGES OF TECHNOLOGICAL SUCCESS: THE AUTOMOBILE

The Design Stage and the Design Constituency

It could be argued that the design stage of the American automobile lasted until roughly 1910. The previous decade and a half was marked by turbulence and uncertainty, hallmarks of the design stage of new technologies generally. In 1900 it would have been difficult to say just what "the automobile" was. Steamers, electrics, and internal-combustion horseless carriages, assembled by a host of small enthusiasts, caught the public's attention as romantic wonderments, but whether any design would achieve the status of a serious and durable transport technology was yet not clear. Bankruptcies were, it seemed, nearly as common as mechanical breakdowns. One big winner, the Ford Motor Company, emerged from the confusion. The 1908 Model T may well be the most successful single design in American history; its durability and reasonable cost fit the social context almost perfectly.

Profits from the enormously popular "T" permitted Ford to invest in a radically new production system, the 1914 moving assembly line. This innovation, more than any other development, created what we can call "the American style" of making automobiles. Instead of bringing individual parts to a fixed station run by a skilled worker, the moving line used semiskilled workers, each having a single task, while the car moved past them. The innovation caused dramatic increases in production efficiency, but the improvement carried a high price tag. Ford transferred the judgment about quality from workers to supervisors, and by that rearrangement of shopfloor politics the company institutionalized a new form of worker alienation that would affect the quality of automobile assembly for the rest of the century.[10]

This shift of adult decision making from worker to supervisor embodies the values of the Ford design constituency, which in turn reflect the ethos of the time. Southern and Eastern European immigrants made up much of the automobile workforce and brought with them all the strangeness of alien lifestyles and foreign languages. Ford's admittedly idiosyncratic paternalism and obsession with control fit nicely into a pervasive national fear of immigrant-generated social chaos.[11]

Thus, American automotive technology took definitive shape, a combination of internal-combustion vehicle and mass-production technology. Its early years were "flexible" in the sense that the design process was strongly influenced by its social setting, particularly as embodied in the world view of the Ford design constituency. We should not push the concept of flexibility too far, losing sight of limiting design constraints. The Ford style depended on and continued a century-long technical tradition that originated in the arms manufacturing practices of the U.S. Ordnance Department.[12] Nevertheless, when compared with the next stage in its evolution, the flexibility of the automotive design stage left the technology remarkably open to contextual influences. In this sense, we can speak of it as embodying some set of values, as having a "style" of its own.

The concept of style is not limited to individual technologies. The people and institutions with access to the venture capital that new technologies always require—in short, the design constituencies of the period—ordinarily hold cultural hegemony in their society. They belong to the group that shapes the dominant values and symbols of society. Although members never form a single historically tidy group, as a technological conspiracy theory might suggest, they tend to view the world from the same perspective. Consequently, we can look for a set of successful technologies that, in any relatively stable era of history, embody the technological style of their society.[13] It is no accident that Ford's moving line, Taylor's scientific management movement, Sperry-designed servomechanisms, tear gas for crowd control, and postwar consumerist advertising techniques reflect similar values: they all use integrated systems to control potentially chaotic exogenous variables.[14] To be sure, technological style does not operate as the sole cause of prevailing cultural values. Values embedded in successful new technologies originate in the world view of those who design and maintain them. Ford's paternalistic obsession with control predated the moving line. On the other hand, technological style fosters as well as reflects values. The host of American adaptations to the automobile exemplifies the way society accommodates itself to the designs of its successful technologies. In the process, values that fit the technology achieve societal momentum, whereas values that do not fit diminish in importance.

The Momentum Stage and the Maintenance Constituency

The truest measure of success for a new technology is found in the array of adaptations to its design constraints. Between 1915 and 1970 America responded to the automobile with a host of institutional creations from motels and suburban architecture to legislation governing highway funding, insurance, and licensing. In the process, millions of individuals and hundreds of institutions coalesced in a little-noticed social grouping that can be called the automobile maintenance constituency. What gave them social coherence, despite their wide economic, political, and cultural diversity, was the fact that they had all come to benefit from and depend on the increasingly popular automobile system. Automotive engineers; workers in automobile, rubber, steel, and glass plants; and proprietors of automobile-dependent small businesses, among others, all depended on automobiles for their livelihood. Ordinary citizens also adopted the automobile as an essential means of transportation. Examples are legion.

The very size and diversity of its maintenance-constituency membership indicates the societal momentum achieved by the automobile. When a technology becomes essential across such a broad spectrum of the body politic, it becomes nearly unthinkable for the technology to change in any substantial fashion. Imagine, for example, that the nation were to decide, for whatever reason, to shift from an automotive to a high-speed rail transport system. From an abstract technical perspective, the change might be conceivable and desirable. However, consideration of the potential social turmoil and economic chaos provides a more accurate reading of automotive momentum and, at the same time, a sense of how the early flexibility of automotive technology's design stage has been replaced by what might be called dynamic rigidity. The technology has shifted roles from a possible social option, open to a host of different futures, to a culture-shaping and highly specified social force. This shift from flexibility toward rigidity marks the movement of the automobile from the design to the momentum stage.[15]

The Senility Stage

As long as the technology's success endures, the fit between technical design and societal context remains tight. Eventually, however, changes occur in the large context. As new political priorities, shifting demographics, changing tastes, ecological transformations, or competing technological actors come to the fore, the sweet fit between context and technology that characterized the momentum stage begins to unravel. Consider a few recent examples that heralded the senility of America's dominant automotive style. The American automobile, a "gas-guzzler" stressing power over fuel efficiency, reflected the very cheap petroleum that marked its context of origin. Beginning in 1973, for a variety of political, economic, and religious reasons, the Organization of Petroleum Exporting Countries drove the price of petroleum to dramatic new heights, and fuel-intensive cars became less attractive. At the same time, the Ford style of automobile production in which workers make no decisions about quality had created cars whose fit and finish suffered accordingly. Americans became used to these standards and accepted them; it was understood that quality control was ultimately handled by the automobile dealerships. In another contextual change, however, the mid-1970s brought new competitors into the car-selling arena. Japanese automotive style, shaped by a radically different world view, put quality-control decisions at the point of production. Japanese quality of fit and finish transformed American tolerance for the earlier Ford style car. Finally, in the mid-1960s people concerned about air quality began to achieve political power. They passed laws that put constraints on yet another dimension of American automobile design.

Unfortunately, the dynamic force that provided the momentum of a once-successful technological style tends to remain rigid even as a new contextual situation begins to demand change. At this point proprietors of the technology might respond in one of two ways. They can attempt a return to the radical flexibility of a new design stage, or they can use their considerable economic and political power to try to force the context back into the sweet fit of the previous momentum stage. General Motors' Saturn project, stressing Japanese-style managerial and production techniques, exemplifies the attempt at renewed flexibility. Instead of trying to transform one of the five established divisions (Chevrolet, Oldsmobile, etc.), GM created a new division for the experiment. When I asked a GM observer why the firm was risking the uncertainties of an entirely new divisional venture, he responded: "A completely fresh start gives us the chance to avoid the entrenched attitudes of much of existing GM management and the mistrust of current labor-management relations in a number of locations."[16] It would be hard to find a more apt example of the tension between flexibility and rigidity at the heart of a powerful technological enterprise. On the other hand, GM's "Mr. Goodwrench" advertising campaign represents an attempt to change the context to fit existing technological style. Mr. Goodwrench's warm and trustworthy demeanor is designed to convince potential buyers that "quality GM parts" mean quality GM automobiles. It is far easier and much less expensive to promote the image of quality than to accomplish it on the shopfloor and in the managerial suite. Whether American automobile firms will achieve sufficient flexibility to respond successfully to the challenges of their changing context remains an open question.

Impact Constituency

Thus far we have seen five of the constituency model's six components. It remains only to note the place of the impact constituency—people and institutions who lose because of the design of the new technology. Most obvious are those who depend on competing technologies. In the case of the automobile, the maintenance constituencies of other forms of transport technology—from livery stable operators and blacksmiths to railroad management and workers—lose because of automobile success. A second type of impact constituency includes those who share the costs of a technology without receiving its benefits. In cities such as Detroit, where tax revenues support freeway networks at the expense of nonautomotive transit systems, residents who cannot afford a car fall into this class. Their loss is compounded if automotive system requirements directly harm them, as when a freeway destroys homes, shops, and walking patterns of the neighborhood in which they live.

Finally, and more subtly, almost everyone who gains from a successful technology loses at the same time. In other words, members of the maintenance constituency are often members of the impact constituency as well. For example, rush-hour drivers suffer from excessive isolation because of the atomistic individualism of the automotive form of mass transit. Suburban architecture, leading as it does to a lifestyle without much walking, leads to empty sidewalks and correspondingly unsafe streets.[17] The automobile, welcomed in the early part of the century as a solution to pervasive urban pollution (horse manure), became a potent source of pollution in its own right. Trade-offs abound; there is no technological "free lunch."[18]

What can we learn from this admittedly sketchy overview of the American-style automobile? Contextual history helps historians to interpret, in specific detail, how Americans have embodied a societal vision in their automotive system and, perhaps more important, how that system, with its inherent biases, influences them in return. The contextual perspective reveals the automobile not as a value free system or an inevitable cultural force, but as the technological incarnation of a series of historically specific choices that are "political" in the root sense of the term.

PART II: STANDARDIZATION REPLACES NEGOTIATION

Automotive history, for all its inherent richness and complexity, cannot be understood in isolation from a larger sociotechnological context. Turning our attention to the nineteenth century, we find evidence that the decisions taken by the automobile's design and maintenance constituencies were part of a much larger social process by which the nation adopted a distinctive technological style. The complete argument is too complex to be presented in this essay, but space does permit an overview of one key aspect of twentieth-century technological style in the United States—namely, the option for standardization at the expense of negotiation.[19]

Until 1870 or so the challenge of conquering the wilderness shaped the dominant technological style in the United States. As generations of men and women from Europe or the more settled eastern United States headed west, their longing for a "middle landscape," a livable place carved out of the wild, grew into a central element of the American character. The land itself—fifteen hundred miles of virgin forest, another thousand of prairie, the Rocky Mountains, the Sierra, and the intervening desert—inspired an American dream. To non-Indian eyes it was empty of history but full of both promise and danger, a manifest destiny challenging the best people had in them. Building a human place, clearing fields and rivers, constructing homes, roads, canals, bridges, tunnels, and cities preoccupied the technological imagination. Americans honored technical expertise as "know-how"—a blend of often crude rules of thumb, occasional engineering elegance, and courage, together with an intimate knowledge of local terrain—as the context whose constraints defined the limits of every project. Technological style, then, demanded a continual negotiation between the skills, tools, and plans of white Americans and the godlike wilderness they sought to conquer.[20]

Despite their passion for freedom and individualism, Americans found negotiation a basic necessity in human interaction as well. People needed one another in an empty land.[21]

Long before this technological style fell from preeminence, however, a successor began to exert its influence. Beginning in 1815 with the U.S. Ordnance Department's commitment to standardized uniformity in weapons production, a new technical ideal gathered momentum in the land.[22] It embodied a radical shift in values. The older style's negotiation with nature and with coworkers shifted toward standardization's precision design and centralized authority. The resulting American factory system created a host of new products, and in the process transformed the relationship between manager and worker from the sometimes respectful and sometimes tumultuous interaction of the early American small shop to heavyhanded enforcement of work rules coupled with the deskilling of workers through increasingly automated machines.[23] The transformation was not limited to the factory. Little by little, a broad range of technological endeavors began to adopt the standardization ideal. Three examples will suggest the flavor of the new style.

Railroads evolved from a turnpike model—state-owned and supported roadbed—to a centrally owned enterprise that included roadbed and most system components. At the same time, the relationship of railroads to their surrounding context changed dramatically. J. L. Larson contrasts the design of grain-shipping facilities for St. Louis and Chicago in 1860. The St. Louis design demanded bagging the grain, loading it on train cars, off-loading it at the outer edge of town where the tracks ended, teamstering it across the city, and loading it again onto river boats. The Chicago design permitted bulk loading on grain cars because the track ran all the way to the docks where it was off-loaded onto grain boats. Larson concludes his description with the following provocative sentence.

> If the Chicago system was a model of integration, speed, and efficiency, the St. Louis market preserved the integrity of each man's transaction and employed a host of small entrepreneurs at every turn—real virtues in ante-bellum America.[24]

The arrangement in St. Louis required negotiation as part of the shipping process, whereas the more capital-intensive Chicago design achieved greater efficiency and permitted railroad management to ship grain without needing to negotiate with that "host of small entrepreneurs at every turn."

The geographical range and managerial complexity of the railroads fostered standardization on many levels. The railroads' need for precise timetables transformed a land of multiple, local times into 1885's single system of standard time divided into our now familiar four time zones. The depot, that little building where town and rail line met, gradually evolved toward a standardized architecture that reflected the triumph of railroad system over local style. The telegraph, essential as a railroad information network, became the vehicle for nationwide, standardized news with the development of the wire services.[25]

We see the same trend in the changing character of electrical systems. If we lived in a small town in 1890 and held a town meeting to decide whether to invest in electricity, our debate could lead to a "yes" or a "no." If "yes," we would shop among the three manufacturing firms—Thompson-Houston, Edison General Electric, and Westinghouse. Once installed, the system would be ours, a tool we had purchased to serve our needs. We townsfolk were independent vis-à-vis the technology. We could take it or leave it, and we could purchase one or another type as well. Today our little town no longer maintains this independent position. From the perspective of electrical systems management, our town serves as a functional component. We no longer "negotiate" with the electrical technology. Our dependence on electricity, together with the complex requirements of centralized generation and transmission systems, has changed the earlier negotiated relationship to one of conformity.[26]

A major redefinition of advertising extends the pattern. Beginning in 1923 with the arrival of Alfred P. Sloan as president of General Motors, the task of marketing new cars shifted from Ford's approach, stressing the economy and technical competence of an unchanging Model T, to fostering cyclic dissatisfaction with one's present car, the basis of "turnover buying." Continued expansion of the mass-production system required turnover buying for, as the recent automobile recession demonstrated, when too many owners hold on to their cars for too long the new-car market stagnates.[27]

Sloan's marketing strategy at GM utilized techniques pioneered by advertising agencies during and immediately after World War I. The new advertising shifted attention from product virtues to consumer benefits and moved away from an earlier "announcement" style toward an evocative rhetoric based on the assumption that consumers were motivated more by emotion than by reason. The advertising revolution fostered a sense of personal inadequacy as the basis for impulse buying. Consumerist marketing does not work with money-conscious shoppers who tenaciously negotiate every purchase or who stay home content with what is already there.

Consumerist advertising marks the epitome of standardization's triumph over negotiation. Whereas the Ordnance Department tried to standardize musket parts and the work patterns of skilled armorers, and centralized railroads later in the century began to standardize equipment and to expand corporate control over previously independent entrepreneurs, the new form of marketing tried to extend standardized conformity into human motivation itself.[28]

These three vignettes indicate a radical shift of the values embedded in our normal technologies. "Technology" once meant "the tools and techniques that humans make and use for their purposes." Twentieth-century style suggests a new definition for our most complex technologies: "elegantly designed systems on which we depend for our survival." Twentieth-century systems tend to resolve the problem of peers negotiating with one another by bringing once-independent negotiators inside the system as functional components.

Three examples do not do justice to the scope of this value shift. I have, for example, hardly mentioned our changing relationship with nature. Thus, our century-long avoidance of waste-disposal problems, seen so vividly today in acid rain and toxic pollutants, reveals a consistent pattern of overriding or ignoring nature's constraints. Only recently have the majority of Americans begun to recognize that nature "has a say" in the technological endeavor.[29]

Standardization admittedly yields substantial benefits. Standardized mass production permits a much larger segment of the population to own relatively high-quality manufactured goods, improving the standard of living for poor as well as rich. Even more important, standardization fosters and rewards the virtue of precision. Living as we do in an age where sophisticated systems such as telephone networks, electric utilities, medical technologies, and computers are commonplace, we find it hard to imagine a world where the art of making steel seemed almost magical. Precision design has joined the family of the elegant arts even as it makes possible the systems on which our communication, our health, and our productivity depend.

With its many virtues, however, standardization carries significant liabilities; most important for us is the atrophy of the ability to negotiate. Negotiation is a messy business. When mutually independent peers must find a common way of proceeding, their different world views, vested interests, and styles make the outcome unpredictable whether the people in question are skilled workers and managers, shoppers and sellers, or independent nations. In the quest for a common good, no one participant's version of the best solution will be adopted. Despite its inefficiencies, the interdependence inherent in negotiation requires and therefore fosters a capacity for engagement with others unlike oneself and an abiding sense of the value of agenda other than one's own.[30]

Of course, negotiation should not replace all conformity. In the form of simple good manners, civic duty, or a host of other pragmatic arrangements, conformity makes ordinary life possible and bearable. If we tried to negotiate each aspect of life at every turn, we would wear ourselves out with endless wrangling. Still the tendency to replace negotiation with standardized conformity creates a serious imbalance in American society. These two societal virtues—standardization's precision and negotiation's uncertainties—seem to work best in creative tension with the other.

The argument, to this point, runs as follows. The American automobile— that is, the car, the Fordist production system, and GM's consumerist marketing style—provides a typical example of the way a successful technology first embodies and later enforces a specific set of values. The case also

,uggests some of the ways that a once-powerful technological style might elate to its context when exogenous changes begin to diminish its momentum. Finally, an overview of America's option for systemic standardization suggests not only that the American automobile is part of a larger and older technological style but also that a diminished ability to negotiate is the most significant trade-off for the benefits of standardization.

The final section rests on a related hypothesis—namely, that the problem of excessive individualism, William Appleman Williams's great American "evasion," stems not from the option for standardized systems, but rather from the atrophy of a complementary habit of negotiation. To summarize in advance, members of the three technological constituencies—design, maintenance, and impact—have the capacity to foster or to evade the precise kinds of political negotiation that would remedy an overly passive and individualistic American style. To foster such negotiation without destroying the possibility of effective system design and operation is, however, a challenge that demands a certain discernment by constituency members. When does masking one's role in the success of a technology foster technological creativity? When, on the other hand, does it evade healthy negotiation? Taking each constituency in turn, let us consider the question.

PART III: ELUSIVE TECHNOLOGICAL CONSTITUENCIES

The Design Constituency

The principle of paleontologist Teilhard de Chardin—"Nothing is so delicate and fugitive by its very nature as a beginning"—holds as much for new technologies as for living species.[31] Both the psychology and the politics at work during the origin of a new technology make it difficult to catch designers in the act. Their penchant for privacy stems in part from the very nature of technological creativity. Thus, although the design *process* involves negotiation to define project goals, the *act* of designing cannot occur in an atmosphere of endless discussion. At some point, debate must cease to make room for concentration on the problems at hand. It is, therefore, perfectly reasonable to say "we are going into hiding so that people can't get at us."[32]

Designers can also choose secrecy for a more questionable reason, namely to exclude some parties from design negotiation. The contrast between Japanese and American corporate styles of new program implementation illustrates the point. The Japanese follow the "Ringi" principle: a new project cannot be approved at a higher managerial level until every affected party on the lower level has studied it, debated it, and finally indicated their approval. At this early stage, American firms tend to take a shortcut. Upper management can approve a project without such broad-based negotiation. Other affected parties in the firm have to adapt to it without having their say at the beginning. There follows a long period of "working the bugs out" that can include, among other things, enforcing companywide conformity on people who are angry at having been excluded from the original design process.[33]

Avoidance of negotiation trades short-run efficiencies for long-term alienation and discontent, thus diminishing rather than enhancing technological creativity. In short, the effectiveness of a technical design must be measured not only in terms of its structural efficiencies but also in terms of the durability of its long-term relationship with the larger social context. The principle is not limited to in-house corporate relationships, however well they exemplify the dynamic in question. At every level of society and in situations of cross-cultural technological transfer as well, technologies are politically effective only to the extent that their designs reflect some basic consensus about the goals that the technology serves.[34] In terms of the three-constituency model under discussion, this need for consensus raises a question about the way maintenance-constituency members participate in technological consensus.

The Maintenance Constituency

Those who benefit from a technology adapt to its constraints and become dependent on it. As we noted in the case of the American automobile, the number and diversity of people and institutions who depend on a technology constitute the measure of its cultural influence. They sustain its momentum precisely because they cannot get along without the particular technology. On its face, this relationship does not seem very elusive at all. Unlike designers, people who benefit from a technology do not try to avoid notice; indeed, those who preside over technological systems and those who use them often take pride in their access to the technology in question.

A closer look at the dependent character of the relationship, however reveals its own pattern of avoidance, once again for a mix of motives. On the one hand, chronic mistrust of technical systems and excessive attention to technological constraints make a recipe for mental illness. Hyperawareness of complex technical systems, or of anything else, immobilizes the human actor. So, for very good reasons, system insiders and system users often ignore their role in maintaining system momentum and their dependence on it.

On the other hand, the maintenance relationship can be repressed because it makes us uncomfortable. If we stopped to think, for example, that the automotive system on which we rely itself depends on a normally durable but not infallible, network of technological support systems and that the entire system has trade-offs for society and personal life, we might become more conscious of the fragility of an essential technological system that we would like. Masking the maintenance relationship can, therefore, take the form of healthy selectivity or evasive passivity.

At the time of this writing, two recent events suggest the political and social complexities of maintenance-constituency evasiveness. In mid-January of 1986 the space probe, *Voyager 2*, approached Uranus almost five years and one billion miles after launch. Its computer control systems timed the encounter within one minute of the original plan while programming a rotation that permitted extraordinary photographs of this distant planetary neighbor. Several days later, before a stunned America, the space shuttle *Challenger* exploded, killing six astronauts and one schoolteacher. *Voyager*'s elegant achievement was greeted with almost blasé acceptance and so, before its launch, was the anticipated success of *Challenger*'s mission.

The gradual unfolding of the NASA *Challenger* investigation provides the public with examples of technological fallibility; in the automotive field, a comparable situation was the series of reports attending the lawsuits that stemmed from the disastrous design of Ford Pinto gastanks.[35] In such cases, the tension between governmental investigators or the damaged party bringing suit, and those who disclose system failures only under duress reveal the common tendency of insiders to cover up the vulnerabilities of their technology. At the same time, the public shift of mood in the *Challenger* disaster—from initial boredom at "another shuttle launch," to shock, grief, and fear, and finally to anger at NASA failure to reveal design flaws—tells us that it is not only insiders who tend to ignore technical fragility. One needs only contrast public nonchalance about *Voyager 2* and preexplosion shuttle missions with the rapt attention that attended the first manned moon landing to see how hard it is for the body politic to sustain attention for the complexities of familiar systems. *Challenger*'s failure and its emotional aftermath demonstrate that taking the system for granted was due not simply to confidence in normally dependable technologies but also to a form of social amnesia that takes the form of willful technological illiteracy.[36]

Such amnesia has serious consequences for civic consciousness. When we habitually suppress awareness of system complexities, we treat the elegant humanity revealed in their engineering with contempt and, at the same time, grant them autonomous, godlike status that dwarfs the human person. What difference can individuals make in a world dominated by massive technologies? Are we anything more than passive drifters on the inevitable technological tide? Even the technically competent suffer these feelings of inadequacy. Outside one's area of expertise, each depends on technologies too complex to comprehend. Still, democracy requires deep confidence in one's creative power. To participate in the life of the body politic, citizens must believe that their judgments carry value in the social equation.[37]

For members of a maintenance constituency, then, "technological literacy" requires more than the simple ability to use a technology. It includes the capacity to interpret major technologies as systems that must always be maintained by someone, that always cost money, and that always have associated trade-offs. Still, if our every waking moment were obsessed with consciousness of the political and technical fallibility of every system, society would disintegrate in anxious chaos. A healthy maintenance-constituency relationship, then, balances moments of contemplation that foster alertness to the full political and technical dimensions of technological systems with longer periods of day-to-day system use.

We turn, finally, to the most difficult political question of all, the relationship between a successful technology and those who lose because of the way it is designed.

The Impact Constituency

Paulo Freire's maxim, "No oppressive order could permit the oppressed to begin to question: 'Why?'" and his groundbreaking *The Pedagogy of the Oppressed,* from which it is taken, illuminate the seemingly mysterious capacity of impact-constituency members to remain invisible.[38] Those who lose because of the design of dominant systems—technological, economic or political—are most often hidden from view. Not surprisingly those who design and preside over successful systems tend to ignore those who lose. Industrial capitalism, in its pure form, explains the poverty of those who lose as a result of their own character deficiencies.[39] Freire's argument, however, while it includes a critique of the laissez-faire capitalist position, calls attention to the oppressed themselves who tend to evade consciousness of the systemic causes of their suffering. His "pedagogy" encourages the struggle to achieve consciousness of systemic oppression as its cornerstone.[40]

Why do the dynamics of the impact constituency generate numbness? One story from my experience with Native Americans shows a common pattern. When I lived with the Weasel Bear family on Pine Ridge in the summer of 1968, we depended on our car, a twelve-year-old Ford coupe, because we had to drive two miles for water. The Ford's trunk was sprung, the back windows did not work, and the transmission was so bad that under five miles an hour the engine died and above fifty miles an hour the car began to shake violently. A hole in the radiator made us carry water for refills every five miles or so.

One day I asked my Indian grandmother Rose, "What did you pay for this thing?" She said: "Five hundred dollars." I asked where she bought it and found that she had bought it from a car dealer in an off-reservation town. Reacting like the lawyer's son that I am, I said: "We've got to do something about this! We should call the Better Business Bureau." "No," she said quietly, "Suppose we got them mad; where would we get cars from?" The powerless tend not to think too hard about how systems punish them. Impotence often breeds numbness as a substitute for pain.

The "anti-Luddite" argument fosters the same mentality. The original Luddites selectively broke those machines that they judged damaging to themselves.[41] More commonly, however, the term applies to those who break all new machines because of their inability to keep up with progress. Critics of successful technologies are frequently subject to the charge: "Another Luddite! You hanker for the days of untreated diseases and outdoor toilets. A romantic." Anti-Luddite rhetoric denies the very existence of impact constituencies, claiming that despite obvious costs in the present, new technologies benefit even current losers because, in the long run, progress will make life better for all or at least for the descendants of all.[42]

Even more subtly, members of the maintenance constituency are themselves members of the impact constituency. The negative trade-offs of technological designs do not simply divide a population into those who win and those who lose. Every human artifact—simple mechanism or complex system—embodies some limited set of values and serves limited purposes. Technologies necessarily constrain us while they help us. Thus, as we noted previously, automobiles give us access to independent mobility even as they tend to isolate us from one another. On reflection, however, it can be seen that the limitations inherent in the act of technological design provide an extraordinary focus for healthy social critique. Every culture writes a profile of its governing values in the benefits and liabilities of its successful technologies. By attention to impact constituency relationships, the discerning citizen reads the signs of the times. Contemplating society through the prism of its normal technologies demands hard emotional work; it is never easy to take a critical stance toward the dominant values of one's culture. Nevertheless, the challenge of life in a highly technologized democracy calls for precisely this form of adult consciousness.

There are, of course, very constructive reasons for ignoring the liabilities of a technology. Too much technological negativity immobilizes the critic in the same way that excessive attention to technological fallibility paralyzes members of the maintenance constituency. Thus, critical awareness can be healthy or counterproductive, and discerning the difference challenges citizens of contemporary society. Submission to the myth of autonomous progress, by saying that all the negative impacts of our technologies are temporary and that eventually everyone will be better off, trades active citizenry for passive drifting.

This is not to say that society will ever free itself from negative impacts. The establishment of human structures necessarily results in some situations being better and others being worse. Increased consciousness about how we relate to our technologies will not lead to a technological "pie in the sky" whereby we eventually reform the world so that there are no more impact constituencies.

CONCLUSION

What, then, makes it worth the effort to sort out the politics of tech-nology—to endure the discomfort and uncertainty that necessarily follow from contemplation of successful technologies and their constituency rela-tionships? Put most simply, contemplation is the price of technological and democratic adulthood. William Appleman Williams's maxim bears repeating, "America's great evasion lies in the manipulation of nature to avoid a con-frontation with the human condition and with the challenge of building a true community." Passive technological drifting encourages a childish citizenry because it rests on the premise that the welfare of the body politic is someone else's business. "Confrontation with the human condition" cannot happen in this technologically complex society without the very adult awareness that someone's designs will be adopted and that someone will make decisions about the allocation of resources and about which lines of research will be pursued and which will be ignored. Technological style, with all its nobility and liabilities, remains the work of human beings.

Thus, the ideology of autonomous progress poses more than a histo-riographical problem for those who would write a contextual history of technology. Seductive and inherently passive, talk of progress alone erodes the fabric of societal life itself. Thus, by choosing to do a *social* history of *technology,* balancing sophistication about design constraints with an equally sophisticated alertness to social forces, historians of technology help to create a language that liberates the individual while encouraging participa-tion in the labor of human community. To run the risk of becoming engaged in the political debate means, when seen from this perspective, learning to sort the tangled strands that are continually woven and rewoven as members of society embody their values and world view in technologies that reflect and then reshape a vision of the common good.

NOTES

1. Cited in James W. Carey and John J. Quirk, "The Mythos of the Electronic Revolution," *The American Scholar* 39 (Spring, Summer 1970): 411. *The American Scholar* does not use source notation and neither I nor Dr. Carey have been able to track the original source of the text. In this essay I will frequently, and reluctantly, follow common practice and use the terms "American" and "America" to refer to the United States and its citizens. Although I am aware of the problem involved in this parochial use of the expressions, I have yet to find an alternative that is not inor-dinately clumsy.

2. Frederick Jackson Turner, *The Frontier in American History* (New York: Henry Holt & Co., 1920), chap. 1, "The Significance of the Frontier in American History" (1893), pp. 1–38. See also Henry Nash Smith, *Virgin Land: The American West as Symbol and Myth* (Cambridge, Mass.: Harvard University Press, 1950); Leo Marx, *The Machine in the Garden: Technology and the Pastoral Ideal in America*

New York: Oxford University Press, 1964); and John F. Kasson, *Civilizing the Machine: Technology and Republican Values in America, 1776–1900* (New York: Penguin, 1976).

3. Robert Bellah, Richard Madsen, William M. Sullivan, Ann Swidler, and Steven M. Tipton, *Habits of the Heart: Individualism and Commitment in American Life* (Berkeley: University of California Press, 1985), p. viii. For a similar sociological analysis, see Parker J. Palmer, *The Company of Strangers: Christians and the Renewal of America's Public Life* (New York: Crossroad Publishing Co., 1985). Thomas C. Cochran notes the same patterns, esp. in chap. 5, "Bureaucracy," of his *Challenges to American Values: Society, Business and Religion* (New York: Oxford University Press, 1985).

4. Carey and Quirk, "Mythos," pp. 420–21. Carey, after the manner of Henry Adams in "The Virgin and the Dynamo," focuses almost exclusive attention on electricity to explain American technological style. I differ slightly, seeing electronic technology as one major strand in a larger cultural pattern of standardized systemic technologies, as discussed in my Part II. Richard Slotkin advances much the same argument in *The Fatal Environment: The Myth of the Frontier in the Age of Industrialization, 1800–1890* (New York: Atheneum, 1985).

5. David Potter, "Abundance and the Turner Thesis," in *History and American Society: Essays of David M. Potter,* ed. Don E. Fehrenbacher (New York: Oxford University Press, 1973), addresses the same point in his critique of the frontier hypothesis as articulated by Turner and Walter Prescott Webb. Potter argues that Turner and Webb attributed too much influence to the ideal of wilderness America and "did not recognize that the attraction of the frontier was simply as the most accessible form of abundance" (p. 127). For Potter, the source of American creativity is the promise of abundance and not simply the bounties of Nature. In so arguing, he overlooks the mythic role of Nature as a "feminine" principle of renewal over against a more "masculine" technology and science. For this distinction and its long roots in Western culture, see Carolyn Merchant, *The Death of Nature: Women, Ecology, and the Scientific Revolution* (New York: Harper and Row, 1980). See also n. 20 below.

6. I base these observations on my study of SHOT's contextual discourse in *Technology's Storytellers: Reweaving the Human Fabric* (Cambridge, Mass.: MIT Press, 1985), chap. 5, and on my more impressionistic sense of recent trends in articles published in *Technology and Culture,* monographs, and papers read at SHOT's annual meetings. In this context it is noteworthy that the 1986 Usher and Dexter Prizes were awarded to studies addressing precisely this sort of "thorny political and ethical" question, namely Donald MacKenzie, "Marx and the Machine," *Technology and Culture* 25 (July 1984): 473–502, and Walter A. McDougall, . . . *the Heavens and the Earth: A Political History of the Space Age* (New York: Basic Books, 1985). For a brief study of SHOT's origins and institutional history, see *Technology's Storytellers,* chap. 1.

7. *Technology's Storytellers,* esp. pp. 95–103, 134–48, and all of chap. 5.

8. "Successful" means something more than functional efficiency. In the present context, the term designates a technology that has achieved so much momentum within its society that the society would have substantial difficulty doing without it. Thus, the American automobile system would be "successful," whereas dental floss would not.

9. *Technology's Storytellers,* pp. 192–201.

10. The exact chronology of Ford's shift from the traditional shopfloor arrangement where supervisory personnel made up 2 percent of the workforce to their 1917 arrangement when the supervisory share had increased to 14.5 percent is not altogether clear. Stephen Meyer finds fragmentary evidence for the traditional style as

late as 1913. See *The Five Dollar Day: Labor Management and Social Control in the Ford Motor Company, 1908–1921* (Albany: State University of New York Press, 1981), pp. 50–51.

11. For details of Ford's labor policies during the period see Meyer, *Five Dollar Day*. For similar assessments of contemporary fear of chaos see John Higham, *Send These to Me: Jews and Other Immigrants in Urban America* (New York: Atheneum, 1975), especially chap. 1, 2, and 6. See also Michael Schudson, *Discovering the News: A Social History of American Newspapers* (New York: Basic Books, 1978), pp. 127–34.

12. For a definitive study of the nineteenth-century tradition, see David Hounshell, *From the American System to Mass Production, 1800–1932* (Baltimore: Johns Hopkins University Press, 1984). For a detailed treatment of the technological supports required in the design stage, see *Technology's Storytellers*, pp. 61–69; see pp. 40–45 for a treatment of invention and pp. 45–50 for development.

13. On cultural hegemony, see T. J. Jackson Lears, "The Concept of Cultural Hegemony: Problems and Possibilities," *The American Historical Review* 90 (June 1985): 567–93. See also *Technology's Storytellers*, pp. 192–201.

14. On Taylorism, see Edwin T. Layton, Jr., *The Revolt of the Engineers* (Cleveland: Case Western University Press, 1971; Baltimore: Johns Hopkins University Press, 1986), pp. 134–53. On Sperry's control style of inventions, see Thomas P. Hughes, *Elmer Sperry* (Baltimore: Johns Hopkins University Press, 1971), pp. 45–46, 283–85. On tear gas, see Daniel P. Jones, "From Military to Civilian Technology: The Introduction of Tear Gas for Civil Riot Control," *Technology and Culture* 19 (April 1978): 151–68. On consumerist advertising styles, see n. 28 below.

15. Historical shifts are notoriously hard to fix in time. Design and momentum stages of a technology typically overlap. The technology begins to achieve societal acceptance and a correlative increase in cash flow before it has reached what hindsight will reveal as its mature form. Inventive activity and developmental changes characteristic of the flexible design stage often mark the early momentum stage. Thus, for example, the mid-1920s marketing strategy of General Motors significantly modified the Ford automotive style more than a decade after it had entered the momentum stage. On General Motors's marketing strategy, see n. 27 below.

16. My source requests anonymity. He also noted that the GM-Toyota joint venture—New United Motor Manufacturing Incorporated (NUMMI)—has adopted a "production team" concept reminiscent in some ways of the early twentieth-century, fixed-work-station style that predated Ford's moving assembly line.

17. See Palmer's *Company of Strangers*, pp. 38–39, 46–48, where he argues that these patterns have fostered political apathy in twentieth-century America.

18. For a brief discussion of the tradeoffs inherent in technological style, see Thomas P. Hughes, "We Get the Technology We Deserve," *American Heritage* 36 (October–November 1985): 65–79.

19. I am presently working on a book-length version of this interpretation. For a still-brief but slightly more complete analysis, see my "United States Technology and Adult Commitment" in the November 1986 issue of *Studies in the Spirituality of Jesuits* (available from The American Assistancy Seminar on Jesuit Spirituality, 3700 W. Pine Blvd., St. Louis, MO 63108). I am grateful to the Seminar for permission to quote part of the following analysis of standardization from that number of *Studies*.

20. Carolyn Merchant's *Death of Nature* traces the gradual shift, in Western Europe, from "Nature" defined as the goddess who sets the rules and boundaries for human enterprise to a still feminine but newly passive entity destined for exploitation and conquest. Roderick Nash, *Wilderness and the American Mind* (New Haven, Conn.: Yale University Press, 1967), studies the complementary theme of "wilderness." He finds its earliest meaning—wild, chaotic, and often evil darkness—shifting

in nineteenth-century America to a romantic and nostalgic source of regeneration. As early Americans encountered the virgin wilderness, they found their dreams of conquest tempered as much by nature's raw power as by the crudity of their tools. For other discussions of the middle landscape, see Marx, *The Machine in the Garden;* Kasson, *Civilizing the Machine;* and Thomas Merton, *Conjectures of a Guilty Bystander* (New York: Doubleday, 1968), pp. 33–39.

21. Lest we idealize the American conquest of nature, we should note that the myth of the middle landscape had a special place for Native Americans and blacks. Native Americans were seen as part of the wilderness, godlike in their ability to live in the forbidding terrain and subhuman at the same time, lacking both culture and history. Blacks, on the other hand, were part of the tools that white Europeans used to conquer the wilderness. Neither image, of course, approaches the self-image of these two peoples.

22. On the drive toward standardization in the Ordnance Department, see Merritt Roe Smith, "Military Enterprise and Innovative Process," in *Military Enterprise and Technological Change: Perspectives on the American Experience,* ed. M. R. Smith (Cambridge, Mass.: MIT Press, 1985).

23. The most helpful single source on nineteenth-century labor-management tensions over work rules remains Herbert Gutman's essay "Work, Culture and Society in Industrializing America," in *Work, Culture and Society in Industrializing America* (New York: Vintage, 1966).

24. J. L. Larson, "A Systems Approach to the History of Technology: An American Railroad Example," paper read at the annual meeting of the Society for the History of Technology, 1982, p. 17.

25. Carlene Stephens, *Inventing Standard Time* (Washington, D.C.: Smithsonian Institution, 1983); Mark Duluk, "The American Railroad Depot as Technological Vernacular Architecture," *Dichotomy* 5 (Spring 1982): 10–15. On the early history of the wire news services, see Daniel J. Czitrom, *Media and the American Mind: From Morse to McLuhan* (Chapel Hill: University of North Carolina Press, 1982), especially chap. 1, " 'Lightning Lines' and the Birth of Modern Communication, 1838–1900"; Richard Schwarzlose, "Harbor News Association: The Formal Origins of the AP," *Journalism Quarterly* 45 (Summer 1968): 253–60; and Robert L. Thompson, *Wiring a Continent: The History of the Telegraph Industry in the United States, 1832–1866* (Princeton: Princeton University Press, 1947).

26. For a definitive study of electrification in the United States, Germany, and Great Britain, see Thomas P. Hughes, *Networks of Power: Electrification in Western Society, 1880–1930* (Baltimore: Johns Hopkins University Press, 1983).

27. For "Sloanism," see Emma Rothschild, *Paradise Lost: The Decline of the Auto-Industrial Age* (New York: Random House, 1973), chap. 2.

28. On the new advertising style, see Michael McMahon, "An American Courtship: Psychologists and Advertising Theory in the Progressive Era," *American Studies* 13 (1972): 5–18; T. J. Jackson Lears, "From Salvation to Self-Realization: Advertising and the Therapeutic Roots of the Consumer Culture, 1880–1930," in *The Culture of Consumption,* eds. R. W. Fox and T. J. J. Lears (New York: Pantheon, 1983), pp. 1–38; Daniel Pope, *The Making of Modern Advertising* (New York: Basic Books, 1983), pp. 227–51; and Roland Marchand, *Advertising the American Dream, Making Way for Modernity, 1920–1940* (Berkeley: University of California Press, 1985), chap. 1, 3, and 5.

29. On early twentieth-century avoidance of evidence indicating ecological damage from technical systems, see John G. Burke, "Wood Pulp, Water Pollution, and Advertising," *Technology and Culture* 20 (January 1979): 181–184.

30. We should distinguish two types of negotiation. The type required for adult commitment rests on the conviction that the negotiating partners seek a single

resolution of their problems. They are "in this thing together" and must seek a working consensus. By contrast, America's labor-management negotiations assume that each side tries to get what it can from the other. Trade-offs between two discrete parties substitute for a single consensus.

31. Teilhard de Chardin, *The Phenonemon of Man* (New York: Harper and Row, 1959), p. 120. The passage continues with an observation familiar to every historian: "Beginnings have an irritating but essential fragility, and one that should be taken to heart by all who occupy themselves with history. It is the same *in every domain:* when anything really new begins to germinate around us, we cannot distinguish it—for the very good reason that it could only be recognized in the light of what it is going to be. Yet, if, when it has reached full growth, we look back to find its starting point, we only find that the starting point itself is now hidden from our view, destroyed or forgotten." (Emphasis in original.)

32. These few lines summarize and dramatically oversimplify the tangled subtleties found in real cases of emerging technology. *Technology's Storytellers,* esp. chap. 2, "Emerging Technology and the Mystery of Creativity," treats the matter in greater detail. See also N. John Habraken, *The Appearance of the Form: Four Essays on the Position Designing Takes Between People and Things* (Cambridge, Mass.: Awater Press, 1985), pp. 135–39, for a provocative discussion of negotiation's complexity during the design process.

33. For a comparison of the Japanese and American styles, see Masaaki Imai, *Kaizen (Ky'zen): The Key to Japan's Competitive Success* (New York: Random House, in press). Studies of labor-management conflict due to lack of negotiation make the same point. See, e.g., Meyer, *Five Dollar Day;* Merritt Roe Smith, *Harpers Ferry Armory and the New Technology: The Challenge of Change* (Ithaca, N.Y.: Cornell University Press, 1977); and David Noble, *Forces of Production* (New York: Alfred A. Knopf, 1983). For a survey of recent literature, see *Technology's Storytellers,* pp. 114–18, 176–77, 189–90.

34. A number of recent best-selling books indicate increased awareness of the difficulty and importance of consensus negotiation. Roger Fisher and William Ury's *Getting to Yes: Negotiating Agreement Without Giving In,* ed. Bruce Patton (New York: Houghton Mifflin, 1981), appears to be the best. For another example, see Herb Cohen, *You Can Negotiate Anything* (New York: Bantam Books, 1980).

35. For a summary of NASA and Morton Thiokol coverups in the *Challenger* project, see David E. Sanger, "How See-No-Evil Doomed Challenger," *The New York Times Business Section,* 29 June 1986. On the Pinto case, see Richard T. De George, "Ethical Responsibilities of Engineers in Large Organizations: The Pinto Case," *Business and Professional Ethics Journal* 1 (Fall 1981): 1–14.

36. To take one more from a host of possible examples, we might recall the many water-saving strategies adopted by northern Californians during the drought of 1977. System failure illuminated the design constraints of water supply. Suddenly, ordinary citizens realized that the act of flushing a toilet connected them with a system of pump stations, reservoirs, sewage disposal units, and a host of other components. After the drought eased, however, water use habits gradually returned to the more ordinary levels of semiconscious dependency. Langdon Winner's discussion of the problem of technological ignorance in contemporary society's situation of "manifest social complexity" provides a helpful general discussion of what I am calling "willful technological amnesia." See his *Autonomous Technology: Technics-out-of-Control as a Theme in Political Thought* (Cambridge, Mass.: MIT Press, 1977), pp. 279–87.

37. The harsh treatment often meted out to "whistle-blowers" is a particularly acute form of such individual intimidation. Bucking the system normally carries a high price tag. In addition to the Morton Thiokol–*Challenger* case noted above, see Rosemary Chalk, "The Miners' Canary," *Bulletin of the Atomic Scientists* 37 (Febru-

ary 1982): 16–22, and the special issue on whistle-blowing and engineers in *Technology and Society* 4 (June 1985): 2–31 for multiple examples.

38. Paulo Freire, *Pedagogy of the Oppressed,* trans. Myra Bergman Ramos (New York: Herder and Heider, 1972), p. 74.

39. Laissez-faire capitalism disposes of technological "losers" with a complex argument. Poverty results from personal character deficiency, and human creativity can only be guaranteed within society at large when the penalty for defective behavior is so severe that it serves as a driving motive for individual and competitive advancement. Thus, in his famous "Gospel of Wealth" essay, Andrew Carnegie rejects charity for the poor precisely because it encourages sloth. "It is not the irreclaimably destitute, shiftless, and worthless which it is truly beneficial or benevolent for the individual [rich man] to attempt to reach and improve." Edward C. Kirkland, ed., *The Gospel of Wealth and Other Timely Essays by Andrew Carnegie* (Cambridge, Mass.: Harvard University Press, 1962), p. 31.

40. Karl Marx's famous aphorism, "religion is the opium of the people," makes the same point. Insofar as organized religion suppresses awareness of systemic oppression, it works to numb the consciousness of the impact constituency. Marx ignored, or perhaps had little contact with, religious traditions that, beginning at least with the Old Testament prophets, foster Freire's type of consciousness. For a striking example, see Herbert Gutman's study of working-class protestantism in late nineteenth-century America, "Protestantism and the American Labor Movement: The Christian Spirit in the Gilded Age," in *Work, Culture, and Society in Industrializing America,* pp. 79–118. The texts he cites are remarkably similar to recent Latin American "Liberation Theology." See, e.g., Jon Sobrino, S. J., *Christology at the Crossroads,* trans. John Drury (Maryknoll, N.Y.: Orbis Books, 1976); and José Miranda, *Marx and the Bible: A Critique of the Philosophy of Oppression,* trans. John Eagleson (Maryknoll, N.Y.: Orbis Books, 1974). For a definitive study of the relationship between Marx and Christianity, see Arthur F. McGovern, *Marxism: An American Christian Perspective* (Maryknoll, N.Y.: Orbis Books, 1980).

41. For the interpretation of the Luddite argument given here, see David F. Noble, "Technology's Politics: Present Tense Technology," *Democracy* 3 (Spring 1983): 8–24; and Adrian J. Randall, "The Philosophy of Luddism: The Case of the West of England Woolen Workers, ca. 1790–1809," *Technology and Culture* 27 (January 1986): 1–17.

42. Of course, laissez-faire capitalism, in its crudest form, would argue that the worthless poor should die, preferably without offspring, so that the race might excrete its unfit members. Thus, for example: "Well, the command 'if any would not work neither should he eat,' is simply a Christian enunciation of that universal law of Nature under which life has reached its present height—the law that a creature not energetic enough to maintain itself must die. . . . Moreover, I admit that the philanthropic are not without their share of responsibility; since, that they may aid the offspring of the unworthy, they disadvantage the offspring of the worthy." Herbert Spencer, *The Man Versus the State* (1892; Caldwell, Idaho: Caxton, 1965), pp. 23–24.

No Passive Victims, No Separate Spheres: A Feminist Perspective on Technology's History

JUDITH A. McGAW

It is highly appropriate to honor Melvin Kranzberg by suggesting where a feminist perspective might direct new research in the history of technology. Unlike John Adams, Mel served as a "founding father" who "remembered the ladies" in at least two important senses. Initially he challenged scholars to examine the subject of women and technological change in history. He did so by articulating clearly the general consensus that antedated the establishment of the Society for the History of Technology (SHOT): that technology had served generally to liberate women—from housework, from unwanted pregnancy, and from confinement to the home. Later, after feminist scholars within SHOT began to examine and question that consensus, Mel provided encouragement by listening to papers, attending meetings of Women in Technological History (WITH), responding to correspondence, and welcoming younger scholars in the hospitality suite. Remembering his generous attention to my earlier reflections on this subject, I offer him my latest thoughts with gratitude.[1]

What new directions for the history of technology are visible from a feminist perspective? Thanks to the growing maturity of scholarship both in the history of technology and in women's history, I can barely begin to answer this question in a brief essay. Because the growing scholarship manifests increasing diversity, it is also important to state at the outset that I offer only *a* feminist perspective—one woman's sense of the most provocative and productive lines of inquiry. In particular, my interests and expertise lead me to focus on the American Industrial Revolution, although I believe that my observations have wider geographic and temporal implications.

Let me begin by making explicit what I mean by a feminist perspective. Fortunately, for me Evelyn Fox Keller's *Reflections on Gender and Science* provides a cogently argued feminist approach to a field very close to our own. Feminism, says Fox Keller, "seeks to enlarge our understanding of the history, philosophy, and sociology of science through the inclusion not only of women and their actual experiences but also of those domains of human experience that have been relegated to women: namely, the personal, the emotional, and the sexual."[2] I concur. Feminist theory does not call on us exclusively, or even primarily, to pay more attention to women in the history of technology. Rather, it urges us to pay more attention to gender: to those ideologies that have attributed certain characteristics to men and others to women. It alerts us that beliefs about sex differences exert so pervasive and profound an influence that we must take account not only of the ways in which gender assumptions have shaped technology historically, but also of the ways in which our gender notions shape the way we write technology's history. This is not to say that we can forget about studying women in technological history. It is to say that we will do a better job of examining women, men, and technology historically if we heighten our consciousness of gender, if we continually recognize that notions of gender have been so intricately interwoven in the fabric of our culture that neither historical actors nor historians have escaped their influence.

There are, I think, three general ways in which we have unwittingly embodied gender constructs in the history of technology so that attentiveness to that ideology might set our scholarship on a new course. First, we have often forgotten that gender is an ideology, not a biological or behavioral reality; we have, that is, often assumed the existence of differences between men and women rather than accept the historian's responsibility to substantiate those differences. Second, we have given sustained attention only to historical notions of feminine gender; we have customarily studied female actors as women and male actors as people. We have done so largely by confining our scholarship within the boundaries of what nineteenth-century Americans called the "separate spheres."[3] Third, in accord with our culture's attribution of activity to men and passivity to women, we have tended to interpret as passive those domains of human experience that have been relegated, rhetorically at least, to women. Rewriting the history of technology so as to minimize gender biases will demand extraordinary effort, but the alternative is even less attractive. Continuing to view the past through the lenses of gender will obscure more of technology's history than we can afford to miss.

The best place to find evidence that gender is so pervasive an ideology that we easily mistake it for a biological or behavioral reality is in scholarship devoted to women in the history of technology. This is true only because studies of women are, unfortunately, unique in paying much attention to gender. Too often they remain disappointingly incomplete in their willingness to question gender as ideology and document gender as reality.

What we know most about is ideology. A generation of feminist scholarship has taught us that, at least as early as the nineteenth century, American rhetoric began attributing certain characteristics, such as nurturance, emotionalism, passivity, purity, and piety to women; and others, such as self-interest, rationality, aggressiveness, and preoccupation with material concerns to men. But we know relatively little about how this rhetoric influenced or reflected the actual behavior of men and women. Nonetheless, scholarship on women and technology has been marred by assumptions that women are more sensitive, spiritual, or nurturant than men; that men are more hard-hearted, violent, or preoccupied with issues of control than women. We have been anxious to question assertions that women are weak and irrational; we need to be equally skeptical of other aspects of gender ideology.[4]

The pervasive treatment of women as passive offers the best evidence that even feminist scholarship has been less than radical in its willingness to question gender ideology. Consider, for example, most of our scholarship to date on women and technological change. Initially, before Ruth Schwartz Cowan's path-breaking essay, "From Virginia Dare to Virginia Slims," most discussions of the subject argued that technology had liberated women: that labor-saving household technology eliminated a great deal of domestic work, while new machines used in factories and offices reduced the strength and skill many jobs required, creating novel opportunities for women in the workforce. By 1982, when I reviewed scholarship in the field, ample research had made clear that these "commonsense" assumptions would not withstand historical scrutiny. Available scholarship documented that social and economic forces encouraged and permitted the continued employment of women in less mechanized and less industrialized jobs despite profound technological change in the workplace. Meanwhile, substantial changes in household technology left the sex, hours, efficiency, and status of the household worker essentially unaltered.[5]

As I noted in 1982, these revisionist findings reflect a prodigious research effort in a new and developing field. I must now note a less heartening fact. Both earlier and recent scholarship have asked essentially the same question: How has technological change affected women? By contrast, much of our best general scholarship has raised a very different question: How has society shaped technology? How have men come to invent, develop, transfer, and use new machines and processes? In sum, the questions we have asked reflect our gender stereotypes: men are active, women are passive. Not surprisingly, then, the conclusions we have drawn essentially reify gender ideology: men are in control, women are victims.

Against this scholarly backdrop, Ruth Schwartz Cowan's *More Work for Mother* stands out in sharp relief. Despite her clear demonstration that household technology did not free women from the curse of Adam, Cowan was not content to tell a tale of passive victims. Rejecting as insufficient her own earlier argument that corporate capitalists and advertisers shaped women's options and decisions, she portrays her housewives as women who made meaningful choices. She constructs a model that compares in graphic, sensuous, and visceral detail the domestic work environments of two great classes of women over several generations. In so doing she enables us to see new technology through the eyes of various female consumers who assessed new modes of housekeeping by comparison with the domestic activities they witnessed as children and by comparison with the domestic practices of the other class, whose accomplishments they envied or whose circumstances they feared. The result is a remarkable, empathetic portrait of innumerable nameless housewives whose work and technological choices gave their families a higher standard of living. And Cowan musters a wealth of evidence to document that a higher standard of living meant more than roast beef on the table and lace curtains in the windows; it meant longer, healthier lives, more regular school attendance, and greater access to white-collar jobs, among other things. Given that a higher standard of living has been a major benefit and selling point of American industrialization both at home and abroad, Cowan's housewives emerge as leading contributors to the success of American technology. They are active shapers, not passive victims.[6]

This is not to say that the American domestic environment is the best of all possible homes or that American industrialization had no dark side. It is only to emphasize that Cowan's work makes a major departure by recognizing that average women make meaningful choices—choices that are fundamental rather than peripheral to the American technological enterprise. Alas, that Cowan's contribution needs to be underscored is itself a measure of our highly selective willingness to question the ideology of gender. Instead of being welcomed, her emphasis on women's choices has drawn criticism because those choices were made within constraints imposed by capitalism and patriarchy.[7] Such criticism misses the point and, more important, misses the opportunity Cowan's work offers for pursuing new directions in the history of technology.

We can recognize one major new direction when we notice that Cowan does not deny that her housewives were limited in their choices by various social and economic constraints, but that she chooses nonetheless to emphasize women's choices. The message for me is that housewives were not so different from other people historians of technology have studied, people such as inventors, entrepreneurs, corporate managers, and skilled workers. These people also faced socially and economically limited choices, but choices we have nonetheless deemed historically significant. The principal difference between housewives and these other people is that housewives were female and the other people were male. If we continue to view the choices of most men as freer than the choices of most women, we need to scrutinize our own notions of gender. If we have been able inadvertently to treat men and women so differently, we need also to reassess our ghettoized approach to the study of women and technology. For the time being, I think we can best enforce more stringent standards of self-scrutiny by studying masculinity as well as femininity, men as well as women.

Ironically, one major intellectual benefit of a scholarship begun as compen-
satory women's history is our recognition that we lack a history of men. In
our field, for example, we have said virtually nothing about historical changes
in the ideology of masculinity. We have rarely been conscious in writing
history that masculine gender imposed constraints on men and influenced
their decisions about and relationship to technology. It will be hard, if not
impossible, to treat the dynamics of gender in a balanced way until we have
studied both halves of that ideology.

Even within the present structure of our scholarship, we can make a start
toward serious gender studies by consistently recognizing and acknowledg-
ing that the male actors who predominated historically in American engineering, business, and manufacturing were men and not merely people. In other
words, we can no longer afford to write the history of technology as though it
were normal to be male and aberrant to be female. Fox Keller's observations
indicate succinctly the risks inherent in such an approach. When her exploration of the relationship between gender and science provoked a request that
she report "just what it was that I have learned about women," she tried to
explain, "It's not women I am learning about so much as men. Even more, it
is science." Nor was this exchange unique. "The widespread assumption that
a study of gender and science could only be a study of women still amazes
me: if women are made rather than born, then surely the same is true of
men."[8] For the history of technology, the liability of assuming men to be
genderless should be equally obvious: to do so is to view technology rather
consistently, albeit unwittingly, through the lens of masculinity; to observe
our principal subject through a glass, darkly. If we would fulfill our mandate
to study the relationship of technology and culture, we must look directly at
the social shaping of those men who made technology and at the manner in
which their socialization narrowed their perceptions of technological choice.

My own research suggests one example of this process. During the first
half of the nineteenth century, mechanization made men's work in the paper
industry increasingly dangerous, yet there is little evidence that male workers voiced objections to the new hazards of the workplace or that male mill
owners considered safer machines especially desirable. During the same
years American notions of masculinity emphasized that real men show courage in the face of danger and bear pain without complaining, notions dissemi-
nated most broadly in celebrations of the nation's most popular new holiday,
the Fourth of July. At the same time, however, mill workers could still
entertain a reasonable hope of escaping the dangers their jobs entailed by
becoming mill owners, a strategy that earlier paper mill workers had also
used to escape the less dramatic risks that traditional paper making involved.[9]

By 1865, faster, multifunction, continuously operated machines had increased the hazards of the workplace. Simultaneously, the increased capital
and training required for successful mill ownership dimmed workers' pros-

pects of eventual ownership. Nonetheless, neither male workers nor male mill owners evinced increased concern over workplace dangers. I suspect that widespread masculine experience on Civil War battlefields and late nineteenth-century glorifications of battlefield bravery made machinery's dangers less culturally visible to men who worked with it at the very moment that its dangers appear in retrospect to grow more visible.[10]

Paper making was, of course, only one industry. Considering American industry more generally, the pervasive relegation of women to less visibly dangerous work, even in the early years of industrialization, and, somewhat later, growing public willingness to regulate the working conditions of women and children, but not those of men, offer evidence that gender ideology probably permitted, and perhaps encouraged, the development of machinery for men's workplaces without regard for its potential dangers. What little we know about men's working conditions in other industries also suggests that the paper industry was not unique in finding dangerous machines designed for continuous operation to be socially acceptable.

That we know little about men's working conditions in most other industries is, for my purpose, much more important. We have, by contrast, learned a great deal about technology's implications for male workers' wages and their control over work. Certainly these are important issues. But they are also issues very much in accord with nineteenth-century beliefs about masculinity. Men, so the rhetoric claimed, had a natural interest in money and power, whereas safety, cleanliness, or time with their families ranked relatively low in their esteem.[11]

I rather doubt that mere coincidence accounts for our attentiveness in studies of male workers to precisely those concerns featured in masculine gender ideology. Nor do I think that we have merely been prisoners of our own gender socialization. If we have looked *through* masculine ideology at the past rather than looking *at* masculine ideology in the past, we have done so principally because of the sources on which we rely. Trained as we are in faithfulness to the voices of our subjects, we have difficulty recognizing historical problems that our historical subjects do not call to our attention.

There are several features of our sources that help render masculinity invisible. As Fox Keller notes, gender ideology identified the personal as feminine. Gender socialization is an intensely personal experience. I am inclined to believe that, as a result, women have historically had far more to say about the "bonds of womanhood" than men have had to say about the burdens of masculinity. Equally important, the nineteenth-century doctrine of separate spheres designated the public arena as masculine, making the public records on which we so often rely not merely public but also masculine; that is, they are far more likely to comment on femininity, which is aberrant in the public sphere, than to comment on masculinity, which is normal. Likewise, the doctrine of separate spheres meant that men and women very rarely worked at the same jobs or even in the same places. For students of technological change, this division of labor means that, in following what seem to be logical lines of scholarly inquiry—examining the development of a particular technology or a particular industry—we inadvertently study one sex or the other, rarely both. Thus, we miss an opportunity to notice masculine gender constraints by contrast.

Although we can certainly be more attentive to evidence of men's sense of gender, we cannot alter their failure to comment explicitly on the constraints of their sphere. But we can cease taking the "separate spheres" as "logical" units of analysis. We need not accept home and work, women's activities and men's labor, as separate simply because Americans chose historically to separate them spatially and rhetorically. To do so is to look at history *through* gender. To look at gender *in* history we need to question whether the gender spheres were in fact separate; and, if not, what social functions this socially constructed separation served.

We can begin to see those links that gender rhetoric denied by studying both the men and the women in industries employing both. At the very least this strategy should help to remind us that masculinity is an issue worthy of attention. In my own case, for example, examining both male and female paper mill workers forced me, after considerable attention to gender influences on women's work, to suspect that men might be subject to similar influences.

It also forced me to think about what relationship the unmechanized work of women bore to the mechanized work of men, a question that deserves more sustained attention, especially from historians of technology. In this respect, the paper industry was relatively representative of industrializing America, where men usually built, operated, maintained, and repaired the sophisticated new technology, while women labored on the unmechanized margins of industry or in unmechanized domestic settings. In the paper industry, I found this sexual division of labor playing an essential role in mechanization. Women's inspection and finishing work helped maintain product quality and, thus, helped assure a continuing market for the burgeoning quantitive output of the new machines. Likewise, the ability of unspecialized female manual laborers to shift from job to job, as demand for various paper mill products fluctuated, helped employers compensate for their increasing dependence on inflexible fixed capital and specialized male workers. Moreover, social agreement that "woman's place is in the home"—meaning that all women theoretically had homes to which they could return, homes supported by husbands or fathers—enabled employers to obtain such essential services cheaply and to lay off female workers when demand flagged or the mills needed to shut down for repairs. These are only a few of the reasons why I consider nineteenth-century gender construction an integral part of American industrialization. Indeed, it may well be "the most ingenious contrivance invented in industrializing America" and, as such, doubly worthy of our attention.[12]

Like paper making, many industries employed both male and female workers. Some of them, like paper making, have heretofore received little historical scrutiny—glass making and nonferrous metal working, for example. Others, such as textile manufacture, have been subject to intense investigation but with disproportionate attention to a single group of workers.

There were also, of course, industries employing workers of only one sex. In these instances we need to begin questioning whether factory walls are appropriate boundaries for our scholarship. Such boundaries tell us more about the social history of industries than about the logic of production; we need to explore both. In paper making it had long been customary to carry out both fabrication and finishing under a single roof. In many industries, however, factories turned out goods suited not for final consumers but only for other manufacturers, who converted the product into goods for consumption. Understanding how the sexual division of labor shaped technological development will often require us to transcend such social divisions of production. How, for example, did the abundance of poorly paid women sewing leather goods shape the technology of and the labor of men in the tanning industry? Likewise, how did the division of clothing manufacture between skilled, largely male tailors and unskilled seamstresses shape and reflect changes in the textile industry?

Moreover, as the case of needlewomen demonstrates, household walls are no more "logical" boundaries for our scholarship than are those of factories. Women who sewed performed essentially the same task whether they worked outside the home for money, within the home for money, or within the home for their families. Indeed, given the remarkably low income such work generated and the irregular employment it offered, we cannot even assume that women who sewed for the market made a greater contribution to their families' economic welfare than did women who sewed exclusively for domestic use. Their case illustrates vividly that nineteenth-century records have skewed our definition of production by confining our attention to goods destined for the market. Our sources have, that is, given us a nineteenth-century "masculine" definition of production. One consequence has been our general assumption that household manufacture disappeared largely because the growth of certain industries, principally textiles, rendered domestic production of those goods unprofitable.[13] Yet while women ceased to spin, they dramatically increased the time they spent sewing, cooking, and baking, albeit for their families rather than for the market. It seems to me that historians of technology can ill afford to regard such activities as manufacturing only when performed by Levi Strauss or Campbells. At the very least we must examine the growth of new forms of domestic production in order to see how manufacturers came later to identify new forms of market production as potentially profitable.

Likewise, to explain the growth of cotton manufacture, meat packing, flour milling, and sugar refining, we will ultimately need to explain women's increased conversion of cloth, meat, flour, and sugar into household linen, clothing, and a radically different family diet. We might, of course, ease our task and continue to rely on public records of production, dismissing household activities as mere consumption. But to do so begs some rather important questions. Can we safely assume, in accord with gender rhetoric, that while manufacturers devised elaborate systems of wages and discipline to make men change their work habits and become more productive, women "naturally" and voluntarily altered their work and increased their output merely out of love? Or do we need to explain the remarkable convergence of labor patterns in the theoretically separate spheres? Similarly, can we take for granted the emergence of a diet heavy in meat, sugar, and refined flour— features that set it apart from the diet of most of the world's population, both historically and in the present? Because parts of this diet are well suited to deliver energy to workers quickly and cheaply, its development in industrial households makes more sense when treated as part of industrialization, not merely as a consumer payoff of industrial production.[14]

Considered more generally, new forms of household labor also help explain how manufacturers could cheaply and quickly recruit large numbers of women to perform the work nineteenth-century industry required of them. In the paper industry, for example, the ease with which large numbers of women learned paper-making tasks, exchanged one task for another, and adjusted to different mills helped confirm the general nineteenth-century argument that women's work was natural, God-given, and unskilled. From the mill owners' perspective the work was justly designated "unskilled" because it required little mill training and no formal education. Nonetheless, the fact that machines could not be built to replicate the combination of manual and visual skills women brought to their jobs and the fact that employers benefited from the low wages such work commanded should caution us not to accept at face value mill owners' assessments of their female employees' skills. Comparison of women's paper mill work and nineteenth-century housework suggests that housework trained virtually all women in the skills paper mills required, making potential workers plentiful and making the work itself appear "natural" for women, rather than skilled.[15]

Briefly, female paper workers, like most women employed in nineteenth-century manufacturing, performed tasks that required care and attentiveness, despite the monotony of the work. Had their attention lapsed frequently, they could not have fulfilled their function: maintaining quality. Yet the work itself had so little inherent interest that constant alertness must have been hard to maintain. Indeed, the only apparent stimulation women experienced on the job came from their simultaneous conversation with other women. It seems likely, then, that conversation not only made the work bearable but also contributed to alert workmanship. One twentieth-century superintendent acknowledged as much when he concluded from an attempt to employ male paper sorters that men could not do the job because they could not work fast enough while talking.[16]

There is little evidence that the ability to talk while performing monotonous work resides in a sex-linked gene, but there is abundant evidence that women, especially nineteenth-century women, acquired the ability through their socially prescribed preparation for housework. For most women, the availability of cheap commercial goods had modified domestic labor so that women spent more time doing repetitive tasks that required precise movements or visual discrimination, including sewing, cleaning, and laundering. Housewives also devoted increased attention to child care and meal preparation, particularly baking, work that required only intermittent physical intervention and was performed while simultaneously carrying on the monotonous tasks that took up most time. Thus, as girls and women mastered housework, they learned to keep their eyes and hands on their work while keeping their minds on the angel food cake in the oven or on the children playing nearby. Women likewise acquired the ability to listen to children and servants and to issue advice, instructions, or reprimands without interrupting the progress of their needles or dust rags.[17]

While more and more women were taking on the new work called housework during the early years of industrialization, changes in men's work made housekeeping skills less likely to receive recognition or reward. Most obviously, men increasingly earned wages or profits for their labors, whereas housewives theoretically labored for love, a form of remuneration that discouraged women from realistically assessing their skills or bargaining for better pay. At the same time, men's labor drew them from the home, so that they had less reason to understand and appreciate women's skills. Men no longer experienced the demands of simultaneously working and supervising their sons. Their long hours in the factory decreased their awareness of their daughters' gradual mastery of domestic labor and of the skills their wives brought to the household. And their growing involvement with mechanization and its emphasis on quantitative increases in productivity no doubt gave them less sensitivity to the qualitative improvements in home life attributable to women's work.[18]

As industrialization proceeded, then, the doctrine of separate spheres both proved enormously useful and received confirmation. The delegation of housework to women prepared most women to do the tasks industry asked of them, made feminine skills so common that increased industrial demand did not create scarcities, and left both men and women ill-prepared to value women's skills. Simultaneously, increased industrial demand for women's finishing and inspection work kept most employed at tasks so similar to domestic activities that women's jobs reinforced the belief in women's distinctive natural abilities. Needless to say, the same argument could be applied more directly to the relationship between housework and the principal nineteenth-century employments of women: domestic service, the needle trades, and laundering. Housework simultaneously trained the workers and made the work appear unskilled. For understanding woman's place in the nineteenth-century workforce, then, identifying the links between the "separate spheres" is crucial.[19]

Transcending the doctrine of separate spheres in our scholarship not only means recognizing that women worked, even in the home, but also means questioning whether men were so inherently preoccupied with work as gender rhetoric claimed. In particular, taking for granted men's "natural" interest in working outside the home to support their families may be as dangerous as our earlier assumption that women worked only for "pin money." As we have done for women, we might fruitfully ask whether men preferred to work themselves rather than remove their children from school; how widowers coped with the added burdens of single parenthood; and what difference marriage made in men's work choices. If such factors shaped men's willingness and ability to work long hours, change jobs, or move in search of work, they also helped determine viable workplace technology. Similarly, we need to explore the emotional and material costs to men when they failed to conform to the socially prescribed role of "reliable breadwinner." Were such sanctions sufficient to drive most men into jobs that promised only monetary rewards? What limits did family concerns place on men's ability to reject work they considered unrewarding and, conversely, how did the domestic production of workmen's wives help shape men's demands on their employers?

Asking similar questions about male mill owners might help us see small-scale entrepreneurs, the men who helped create the emerging middle class, as believable human beings, rather than as the profit-maximizing automatons too often conjured up by scholars. For example, was a mill owner's high rate of reinvestment reflected in his household? Did his wife compensate with her labor for money that went into the business rather than into domestic tools, new furnishings, servants' wages, or more spacious living quarters? Likewise, how was growing affluence reflected in mill owners' domestic investments? Did it serve principally to promote domestic entertainment, contributing to their continued success by helping them solidify alliances and maintain and extend their access to good, timely business and technical information? In sum, what did credit reporters signal when, along with "honest," "hard-working," "reliable," and "temperate," they listed "married" as a positive attribute in their evaluations of nineteenth-century manufacturers?

Whatever our answers to these specific questions, the larger enterprise to which they might contribute—understanding how the notion of separate spheres fostered American industrialization—is hardly a radical project to propose to historians of technology. The earliest study to promulgate the term "division of labor" in the English-speaking world, Adam Smith's 1776

Inquiry into the Nature and Causes of the Wealth of Nations, argued that the division of labor fostered technological change. More than two centuries later, historians of technology have had much to say about technology's relationship to the division of labor, although, unlike Smith, we have tended to emphasize how technology shapes the division of labor, rather than analyzing how the division of labor prompts technological change. Considering the close ties between technological change and the division of labor, is it not dangerous to our scholarly endeavor that we do not know how changes in our most basic division of labor, the sexual division of labor, shaped and were shaped by technological change in the Industrial Revolution?[20]

Thus far I have argued that the nineteenth-century sexual division of labor did, in fact, abet American industrialization in a variety of ways. It supplied cheap female labor to perform manual operations that maintained product quality and skilled male laborers willing to work with dangerous machinery. It gave men incentives to labor and women training in skills industry required, skills that employers could call "unskilled." It allowed male mill owners, male workers, and machines to process goods only partially, while assuring that women at home and in unmechanized occupations would convert these goods into the components of a higher standard of living. In sum, the sexual division of labor promoted mechanization by limiting the skills machines needed to replicate and by lowering the wages at which both men and women could be recruited for industrial labor. How was it, then, that this highly functional division of labor came to coexist with the Industrial Revolution it supported?

Unfortunately, given our longstanding presumption that domains associated with women are passive, most of our scholarship has readily assumed that domestic activities suited to an industrial society must have emerged as a response to changes in manufacturing. After all, men act; women react. In addition, our tendency to confuse gender rhetoric with social reality has encouraged us to believe that new domestic activities emerged only in the nineteenth century, when the cult of domesticity and the doctrine of separate spheres gave housework literary prominence. For our purposes—assessing where historians of technology should direct their research efforts—it is most important to note that our neglect of technology's eighteenth-century history means that we have identified new domestic roles as nineteenth-century creations largely by default.

In fact, the work of scholars outside our field suggests that Adam Smith's insight, that prior division of labor paves the way for technological change, more accurately describes the relationship between the sexual division of labor and the American Industrial Revolution. Judging from what we know about eighteenth-century household activities, changes in women's roles helped raise the American standard of living long before Samuel Slater set up shop in Pawtucket. Notwithstanding Alexander Hamilton's belief that manufactures would render women "useful," Americans had good reason to create an industrial system predicated on the assumption that woman's place was in the home.[21]

Even in rural areas women's domestic production increased in the eighteenth century. For example, Sarah McMahon has discovered that, at least in New England, late eighteenth-century farm families could count on a more substantial and varied diet, one that embodied increased time devoted to feminine agricultural activities: dairy processing, food preservation, and gardening. The changing composition of New England diet also meant that women spent more hours preparing meals, for the one-pot monotonously uniform bill of fare had become a thing of the past long before the nineteenth century. McMahon's findings dovetail nicely with Carole Shammas's conclusions from household inventories: during the eighteenth century, ordinary households came to own enough domestic furnishings, including eating and drinking utensils, that family meals in the modern sense became a possibility. How men's roles were changing during these years is less clear, but, at least in southern and eastern New England, declining farm size and increasing purchase of goods for the home suggest part-time employment away from the homestead.[22]

Eighteenth-century urban domestic labor reveals the preindustrial emergence of the "modern" sexual division of labor even more clearly. Mary Beth Norton finds urban families purchasing most of their meat and dairy products, freeing men from their traditional roles as feed-crop cultivators, large livestock tenders, and butchers.[23] The market also exempted urban colonial husbands from the traditional work of grain cultivation, harvesting, trips to the grist mill, and seasonal lumbering to lay in fuel. Moreover, most implements and leather goods that rural men had traditionally made and repaired could be purchased in urban craftsmen's establishments. In sum, even before the Industrial Revolution most men's household tasks had already been taken over by specialists in America's most economically developed enclaves. This late colonial era specialization ultimately provided an environment conducive to experimentation with new mechanical contrivances and manufacturing processes.

Meanwhile, urban women (and, probably, many small-town women) no longer had to spin, churn butter, make cheese, and brew beer, among other things. But this did not render them idle and decorative, any more than household technology delivered twentieth-century housewives to lives of leisure. Rather, by the time of the Revolution urban women had "adopted standards of cleanliness for their homes, clothes, and beds that were utterly alien to farm wives," standards that reflected not only increased cleaning and laundering but also marked increases in sewing to supply the sheets, towels, and clothes that helped make greater cleanliness possible. They also performed more mending and altering, making clothes more comfortable, more stylish, and warmer in winter. Simultaneously, large and regular urban markets supplied a more varied diet, including more fresh meat and vegetables, which, as in the case of New England farm wives, entailed more frequent and complex meal preparation.[24] At least some eighteenth-century urban women also came to live in larger dwellings with more rooms, more storage spaces, and more furnishings. Such accommodations must have made house cleaning more time consuming by giving women more areas and items to clean.[25] And, as Cowan observes for the twentieth century, specialized rooms, differentiated furnishings, and storage facilities helped raise standards of cleanliness; multipurpose rooms without storage spaces and with their few chairs and single table constantly in use precluded the straightening up that necessarily precedes thorough cleaning. Finally, judging from Shammas's study, women had also assumed most of the work of shopping by the late eighteenth century, even in relatively small towns.[26]

Although the information currently at our command remains sketchy, it indicates our need to investigate the early American origins of our industrial society. It certainly appears that most of the activities that Cowan depicts as making woman's domestic work essential in industrializing America were already central to her role long before industrialization. If so, the sexual division of labor began raising the American standard of living well before technological change in industry sought to achieve the same goal. Perhaps, then, the celebrated nineteenth-century role of woman was not an artifact of industrialization, but rather an eighteenth-century construction so valued that industrialization developed so as to preserve and support it.[27] And, if eighteenth-century Americans had already committed themselves through changes in their domestic arrangements to achieving a higher standard of living, the wholesale commitment of nineteenth-century Americans to achieving the same goal through changes in their manufacturing arrangements becomes easier to understand. In sum, scholarship on eighteenth-century America, including both its homes and its other workplaces, should help us answer the most fundamental question we can ask about industrial technology: not, How did Americans design more productive technology? but, How did Americans come to believe increased productivity was both possible and desirable?

Evidently, then, the home, a major domain associated with women, played an active role in the history of technology. As Fox Keller's work suggests, a feminist approach to the history of technology calls for exploration of the influence on technology of the many other domains of human experience relegated to women. Three of these domains seem to me especially promising: consumption, nurturance, and piety.

Consumption offers the most obvious case in which notions of gender have biased our approach to the history of technology. Although we readily admit that production and consumption, supply and demand, are dynamically interdependent, we have, in fact, written only the production side of the story: a supply-side history of technology. Thanks to Ronald Reagan, I need not belabor the point that such an approach has both political and social implications. What I need to underscore is that portraying production as active and consumption as passive distorts our treatment of the history of technology. In particular, it makes us more attentive to the quantitative than to the qualitative aspects of technological change. We often argue, for example, that, in contrast to the British market, the American market willingly accepted goods of inferior quality so long as they were cheap and abundant. Although it may be true that a belief that "more is better" has deep roots in our culture, we need, at the very least, to trace that notion to its colonial origins.

We need also to recognize that quality is relative, not absolute. In the paper industry customers had a highly developed sense of acceptable paper. Only very gradually did they come to accept intertwined cellulose fiber, rather than intertwined rag fiber as constituting paper, despite substantial differences in the cost and abundance of wood and rags. In this case, at least, changes in production technology went hand in hand with changes in who bought paper and how they intended to use it.[28] Customer interest in quality may have exerted less influence in other industries, but such differences in consumer influence also need to be part of the history of technology.

In proposing more attention to consumption, it may be well to add a few words of caution. It will not suffice to cut corners and do a supply-side history of demand—to study advertising and marketing and presume that consumers were its passive victims or respondents. Cowan's work demonstrates that housewives had their own motives for purchasing domestic appliances, despite the apparent correlation between appliance advertising and appliance purchases. From an anthropological perspective, Sidney Mintz's remarkable discussion of the dramatic rise in British sugar consumption exemplifies the importance of avoiding several other unwarranted assumptions. We cannot presume human definitions of "good taste" to be innate, even when the taste is sweet. He also argues persuasively that merely finding that the upper class acquired something before its acquisition by the lower orders does not suffice to prove that poor people were motivated by a desire to emulate their betters. Emulation, as we already know from the work of Brooke Hindle, needs to be studied rather than assumed.[29]

The study of nurturance seems to me equally deserving of attention from historians of technology. Although we frequently allude to the importance in an industrializing society of what economists infelicitously term "human capital," we have made almost no headway in addressing the issue. Given the rather tenuous connection between literacy and the skills required to invent and operate early American machinery, it strikes me that further examination of literacy rates and school attendance figures will not suffice. Studies of numeracy, building on the pioneering work of Patricia Cline Cohen, may be more to the point.[30] Because proponents of industrialization customarily argued its merits in quantitative terms, numeracy might at least afford a sense of how widely their arguments could be appreciated.

What strikes me as even more deserving of attention is the remarkable series of changes in American childrearing that began in the late eighteenth century, that is, on the eve of American industrialization. During these years increasing numbers of women began to give childrearing more time and attention.[31] At the same time, more men began to work away from the home. In sum, two fundamental features of the modern family grew more common just before the onset of our Industrial Revolution. Can this be mere coincidence?

I think not, in part because these changes in childrearing were linked to changes in what children were taught. Most obviously, the feminization of childrearing meant that boys, who entered rapidly changing labor markets throughout the nineteenth century, were less subject to paternal influence at the very time that simple replication of one's father's career grew less socially desirable. At the same time the virtues eighteenth-century mothers were newly urged to teach their sons included "industry, frugality, temperance, [and] moderation."[32] Although these virtues derive from eighteenth-century republican thought, from which they entered the childrearing literature, students of nineteenth-century manufacturing will recognize them as among the litany of virtues reiterated in credit reports on mill owners and in letters of recommendation that manufacturers supplied to men they considered "good" workers.

We might be suspicious of claims that motherly precept sufficed to inculcate these virtues in their sons, except that we also know that these mothers performed new sorts of housework well suited to teach these virtues by example, a far more potent form of instruction. Although sons did not learn women's tasks, as women took on primary responsibility for child care, sons did spend more hours watching women doing domestic work. Women had some choice as to how much cleaning, laundering, and sewing they performed, so that women who adopted new domestic standards presented models of industry. Similarly, housewives could display frugality in their shopping, temperance in the beverages they consumed and served, and moderation in their dress and disciplinary behavior. Examining feminine nurturance may be especially promising, but we also need to learn about what fathers did when they arrived home. Living in an era when debates over working mothers make a strong case for "quality time," we cannot dismiss paternal influence simply because fathers were less often present. Businessmen's relationships to their sons especially command our attention, for the success of nineteenth-century businesses that were typically family enterprises frequently hinged on the successful socialization of the next generation.

Like consumption and nurturance, piety has generally been viewed as a passive, feminine activity. And, like consumption and nurturance, we have given it short shrift. Yet, at least as late as the mid-nineteenth century, the church exerted a far deeper and more regular influence on the lives of most

Americans than did the state. It says something about our biases, then, that we have devoted far less attention to the former than to the latter. Moreover, one of the few studies to generate widespread interest in religion among historians of technology, Anthony F. C. Wallace's *Rockdale,* treats revivalism as responding to, rather than shaping, technological change. Like nineteenth-century discussions of the home, Wallace depicts the church as a place where women helped people adjust to technological change. Wallace may be correct about religion in Rockdale, although in the Philadelphia area, as in much of the industrializing northeast, revivalism antedated the onset of mechanized manufacturing. In Berkshire County, Massachusetts, at least, I found revivals paving the way for technological change by encouraging new modes of behavior among both workers and mill owners.[33]

Similarly, we have customarily viewed the unprecedented emergence of atheism and agnosticism in late nineteenth- and twentieth-century America as a simple reaction to scientific and technological change: modern science, technology, and medicine rendered belief in, and dependence on, God unnecessary. Such simplistic functionalism certainly errs in underestimating the complex place of religion in men's and women's lives. It also falls short as explanation because it assumes what needs to be proved: that science, technology, and medicine offered and were accepted as superior explanations and modes of control. By contrast, James Turner has argued that liberal Protestant leaders, in their attempt to make religion conform to the emerging scientific and technological order, managed to undermine faith far more effectively than did Darwin or material plenty. Although Turner confines his analysis to a small group of intellectuals, his thesis is one worth testing at the level of mill-town churches.[34]

Finally, a word about methodology. Most of the issues I have raised cannot be addressed solely, or even primarily, through quantitative analysis. Qualitative concerns, humanistic concerns figure prominently throughout. Those of us who have developed social scientific skills will not find them useless, but we will need to rely more heavily upon the humanistic traditions of history, on those methods suited to analysis of texts and of individual behavior and motives. We will also need to divest ourselves of social scientific prejudices against such methods.

Let me acknowledge in closing that contemplating the new, uncharted terrain that a feminist perspective opens to our view can be disconcerting, if not downright terrifying. Doing the history of passive victims and separate spheres is obviously safer and easier; writing about stock characters requires less imagination. But such an approach may also be costly. Among other things, it may help to perpetuate the relatively limited readership our work attracts among nonspecialists, noted recently by Merritt Roe Smith.[35] Although it will take courage to enter unexplored country, I believe it is an essential step toward the goal Mel helped set for us: making the history of technology as central to our history as technology is to our culture.

NOTES

I wish to thank Jeanne Boydston, Tom Broman, Mary Kelley, Alejandra Laszlo, Lynn Nyhart, Eric Schatzberg, and members of the 1984–85 Transformation of Philadelphia Project for helpful comments on earlier versions of this work. I also thank the West Virginia Humanities Seminar and Susan Douglas (Chairwoman, 1985 SHOT Program Committee) for providing the initial encouragement to develop arguments presented here.

1. Adams's response to his wife's plea that he "remember the ladies" in writing the Constitution was: "As to your extraordinary code of laws, I cannot but laugh." Carol V. R. George, ed., *"Remember the Ladies": New Perspectives on Women in American History* (Syracuse, N.Y.: Syracuse University Press, 1975), p. 1.

2. Evelyn Fox Keller, *Reflections on Gender and Science* (New Haven, Conn.: Yale University Press, 1985), p. 9.

3. Briefly stated, the doctrine of separate spheres linked distinctions between appropriate male and female behavior to assumptions about men's and women's essential character and to sex-appropriate spheres of activity. It specified that woman's place was in the home and man's place was at work and in the public arena. The only acceptable feminine activities derived from woman's natural roles as wife and mother, so that housework, teaching, nursing, and other permissable women's employments were viewed as acts of love and nurture, rather than vital economic contributions. In sum, the term "working woman" became an oxymoron. The literature on the separate spheres is extensive, but see Nancy F. Cott, *The Bonds of Womanhood: "Woman's Sphere" in New England, 1780–1835* (New Haven, Conn.: Yale University Press, 1977), for a fine discussion of early nineteenth-century developments.

4. For a discussion of this problem that is also sensitive to some of its roots, see John M. Staudenmaier's review of Joan Rothschild, ed., *Machina ex Dea: Feminist Perspectives on Technology* (New York: Pergamon, 1983) in *Technology and Culture* 26 (April 1985): 283–87.

5. Ruth Schwartz Cowan, "From Virginia Dare to Virginia Slims: Women and Technology in American Life," *Technology and Culture* 20 (January 1979): 51–63; Judith A. McGaw, "Women and the History of American Technology," *Signs: Journal of Women in Culture and Society* 7 (Summer 1982): 798–828. Hereinafter, unless otherwise indicated, my generalizations about women's work derive from the literature reviewed in this essay.

6. Ruth Schwartz Cowan, *More Work for Mother: The Ironies of Household Technology from the Open Hearth to the Microwave* (New York: Basic Books, 1983).

7. See, e.g., Sally Hacker's review of Cowan in *Technology and Culture* 26 (April 1985): 291–93, and Nona Glazer's review, cited by Hacker.

8. Fox Keller, *Reflections on Gender and Science,* p. 3.

9. Judith A. McGaw, *Most Wonderful Machine: Mechanization and Social Change in Berkshire Paper Making, 1801–1885* (Princeton: Princeton University Press, 1987), pp. 309–11, 322–24, and passim.

10. Ibid.

11. There are, of course, exceptions to this generalization. One outstanding discussion of men's workplace hazards is Walter Licht, *Working for the Railroad: The*

Organization of Work in the Nineteenth Century (Princeton: Princeton University Press, 1983), pp. 164–213.

12. McGaw, *Most Wonderful Machine*, pp. 346–55, 373, and passim.

13. Rolla Milton Tryon's classic study *Household Manufactures in the United States, 1640–1860* (Chicago: University of Chicago Press, 1917) still serves as the basis for most discussions of household manufacture for the market. It is worth noting, then, that Tryon's sources, principally federal agricultural census data, tell us little beyond the fact that household textile manufacture virtually disappeared. See McGaw, "Women and the History of American Technology," p. 813 n.

14. For a careful and insightful analysis both of the increases in women's domestic work during the early years of industrialization and of the cultural and economic changes that made it less likely to be regarded as economically productive, see Jeanne Boydston, "Home and Work: The Industrialization of Housework in the Northeastern United States from the Colonial Period to the Civil War" (Ph.D. diss., Yale University, 1984); for a provocative discussion of some of these dietary changes, see Sidney W. Mintz, *Sweetness and Power: The Place of Sugar in Modern History* (New York: Penguin, 1985).

15. McGaw, *Most Wonderful Machine*, p. 353.

16. Ibid.

17. Ibid., pp. 353–54.

18. Ibid., p. 355.

19. Ibid.

20. Adam Smith, *An Inquiry into the Nature and Causes of the Wealth of Nations*, ed. Edwin Cannan (Chicago: University of Chicago Press, 1976), pp. 7 n., 13–14.

21. For an excellent discussion of the denigration of women's nonmarket labor by industrial promoters Alexander Hamilton and Tench Coxe, see Boydston, "Home and Work," pp. 105–9.

22. Sarah F. McMahon, "A Comfortable Subsistence: The Changing Composition of Diet in Rural New England, 1620–1840," *William and Mary Quarterly* 42 (January 1986): 26–65; Carole Shammas, "The Domestic Environment in Early Modern England and America," *Journal of Social History* 14 (Fall 1980): 3–24.

23. Mary Beth Norton, *Liberty's Daughters: The Revolutionary Experience of American Women, 1750–1800* (Boston: Little, Brown and Company, 1980), p. 22.

24. Ibid., pp. 22–23.

25. Gwendolyn Wright, *Building the Dream: A Social History of Housing in America* (New York: Pantheon, 1981), pp. 15–16, 34–37.

26. Carole Shammas, "Consumer Behavior in Colonial America," *Social Science History* 6 (Winter 1982): 67–88; Idem., "How Self-Sufficient Was Early America?" *Journal of Interdisciplinary History* 13 (Autumn 1982): 247–72.

27. This is not to say that Americans valued housework economically. Indeed, Jeanne Boydston ("Home and Work," p. 30) makes a persuasive case that "perhaps as early as the middle of the seventeenth century, and certainly long before any important changes had occurred in the nature of the labor, the public recognition of the material worth of housework began to fade. Thus, an analysis of colonial housework becomes important . . . because it suggests that transformations in the cultural visibility of women's domestic labor *preceded* industrialization (and were not a result of it)." In sum, we may need to trace the origins of the nineteenth-century sexual division of labor back much farther than the eighteenth century, and we will need to transcend the early American version of the separate spheres ideology in order to recognize early American women's labor.

28. McGaw, *Most Wonderful Machine*, esp. pp. 191–206.

29. Mintz, *Sweetness and Power;* Brooke Hindle, *Emulation and Invention* (New York: New York University Press, 1981).

30. Patricia Cline Cohen, *A Calculating People: The Spread of Numeracy in Early America* (Chicago: University of Chicago Press, 1982).

31. Norton, *Liberty's Daughters,* p. 39, and passim. As Boydston ("Home and Work," p. 99) notes, the increased emphasis on childrearing had a material basis in the fact that in 1800 34.6 percent of the American population was nine years old or younger.

32. Norton, *Liberty's Daughters,* p. 243.

33. Anthony F. C. Wallace, *Rockdale: The Growth of an American Village in the Early Industrial Revolution* (New York: Alfred A. Knopf, 1978); McGaw, *Most Wonderful Machine,* pp. 81–88.

34. James Turner, *Without God, Without Creed: The Origins of Unbelief in America* (Baltimore: Johns Hopkins University Press, 1985).

35. Merritt Roe Smith, "Social Processes and Technological Change," *Reviews in American History* 13 (June 1985): 157–66.

From Sex to Gender in the History of Technology

JOAN ROTHSCHILD

WHY *GENDER* AND TECHNOLOGY?

When feminist perspectives began to be applied to technology studies more than a decade ago, the body of work was known as "*women* and technology." More recently, the term "*gender* and technology" has come into more frequent use. Why has this language shift come about? Can we find any significance for the study of technology? In a climate of Orwellian euphemisms—as when Mission Control calls a fatal space shuttle explosion a *malfunction*—we might suspect obfuscation and deception: *gender* appears a more neutral term than *women,* its use designed to mask the fact that such work rests on a feminist analysis. In this manner, those who shy away from a focus on women and from the label feminist might have their fears allayed and be drawn into the new field of research. I do not believe that this is the case. Rather, the language change underscores the growing strength of the feminist approach, reflecting in turn the growth and coming of age of women's studies. For technology, this means an expansion of feminist perspectives, and with it an enlarging and enriching of the discipline. The shift to gender means that feminist perspectives on technology are something more than restoring women to their full place in the history of technology, even as this continues to be a major aim and endeavor. The change from women to gender means that gender as a social construct, signifying feminine and masculine, has emerged as category of analysis for technological research.

Like the effect of light refracting through a prism, the introduction of a new mode of analysis and new perspectives to a discipline can reveal a spectrum of possibilities that enrich the whole. Gender analysis has this potential for the study of technology. The impact is in six areas: (1) gender analysis opens up class and race issues, as well as those of gender; (2) it extends inquiry cross-culturally; (3) it broadens the subject matter of technology; (4) it challenges and modifies theories of technological development; (5) it questions and transforms concepts of technological change; (6) a focus on gender raises new issues and concerns for the philosophy of technology.

What does the term *gender* mean? Feminist theory makes an important distinction between *gender* and *sex*. *Sex* is a descriptive category used to designate female and male. It refers to biological characteristics by which we identify someone as a girl or a boy, a woman or a man. *Gender,* by contrast, is a social category. Through what anthropologist Gayle Rubin has called the development of the "sex-gender system,"[1] the biological sexual division of the species is transformed into the gender categories of femininity and masculinity. Our notions of what is feminine and what is masculine are socially constructed and are reflected in behaviors, beliefs, and social practice and organization. Sex and gender are confounded and merged as social attributes of gender become inextricably linked with biological characteristics of sex, and we ascribe socially constructed "feminine" characteristics to women and socially constructed "masculine" characteristics to men, along with activities each sex pursues. Thus, we have come to view technology as masculine and inherently or naturally male, and have conversely assumed that technology is neither feminine nor the province of females. When gender becomes a category of analysis, it enables us to cut through these assumptions and to understand the ways gender constructs have permeated and distorted our thought and practice.

As gender becomes a category of analysis in technology studies, it also transcends its boundaries to broaden and transform the discipline. Drawing examples from the literature, I will illustrate how this occurs in the six areas mentioned above. First, let us briefly review historiographical milestones in the development of feminist perspectives on technology.

MILESTONES IN THE DEVELOPMENT OF FEMINIST PERSPECTIVES

The first area in which gender issues began to be raised was in the female-associated occupation of housework. Starting as early as 1965 in the Spring issue of *Technology and Culture,* Alison Ravetz introduced the neglected subject in her comments on an earlier article by Peter Drucker.[2] Bringing home Drucker's discussion of how social factors contribute to "technological lag" among ancient occupations in developing countries, Ravetz noted that today, in the developed world, the "ancient occupation" of housework suffers a similar lag. Despite new technology, she wrote, in the past fifty years in England social factors have prevented changes in women's roles and living patterns.

In the early 1970s, Ruth Schwartz Cowan took up the housework theme, asking a series of searching questions about developments in household technology and their assumed effects on women's lives. Her work in this area was initially presented in 1973 at the Berkshire Conference of Women Historians and at the annual meeting of the Society for the History of Technology

(SHOT).[3] Her influential article, "The 'Industrial Revolution' in the Home," which challenged assumptions about both the interaction of technology and women's work and the character of technological change, was published in *Technology and Culture* in 1976.[4] Bringing together more than a decade of her research in this area, Cowan's book, *More Work for Mother,* appeared in 1983.[5]

By the mid-1970s a gender focus had extended to other areas of technology and work traditionally performed by women. At the annual meeting of SHOT in 1976, Cowan moderated a panel session that centered on women and their special relationship to changing technology. The panel featured papers by Daryl Hafter on the effects of the introduction of the Jacquard loom on the "drawgirls" in eighteenth-century France, by Susan Levine on the impact of machine power on female and male labor in the American carpet industry in the nineteenth century, and by Martha Moore Trescott on the significant, previously undocumented, influence of Julia Hall on her brother's discovery of the process for the electrolytic production of aluminum.[6] These papers, along with others produced in this period, were subsequently published in Martha Trescott's edited volume, *Dynamos and Virgins Revisited,* in 1979.[7] The rich outpouring of research, especially in the area of women and work in the history of American technology, was documented in Judith McGaw's review article published in *Signs* in 1982.[8]

Events were moving organizationally as well. Women in Technological History (WITH) was established as an interest group within SHOT in 1976.[9] Speaking to sex-gender issues for both researchers and research, WITH was formed with two kinds of scholars in mind: those applying gender analysis to technology, and women working in the technology field, regardless of the subject of their research. In this way, WITH accomplished two aims of a focus on gender. First, it recognized feminist research as a growing and legitimate area of technology studies, whether engaged in by women or by men (WITH membership has always been open to men). Second, by recognizing that women *were* and *are* involved in technology research, it began to dispel the myth that technology was an exclusively male activity and practice. In this same period, in response to pressure, women began to be appointed to SHOT's advisory and executive councils, and as advisory editors for SHOT's journal *Technology and Culture.*[10]

The growing gender activity in technology reflected the rapid development of women's studies in the social sciences and the humanities, which had begun in the late 1960s. Although science and technology took longer to draw feminist attention, gender perspectives in the history of technology gained impetus as a result of history having been one of the first disciplines to develop strong feminist scholarship. The work cited above is evidence of this. Martha Trescott's anthology brought together important research on women and technological history, opening with Cowan's key essay on the differences between female and male experience of technological change, and using the article's research categories in part to structure the book: women operatives in industry; women as inventors, engineers, and scientists; women as house-wives and homemakers; and females as children and bearers and rearers of children.[11] McGaw's comprehensive review essay in *Signs* three years later charted a strong development of feminist research in two areas of the history of American technology—technology and women's work outside the home, and technology and domestic work—while noting the lack of research (at the time) in such areas as reproduction and women inventors, and indicating that much of the field was still to be developed.[12]

My own edited collection, *Machina Ex Dea,* published in 1983, sought to cut a wider disciplinary swath, with the original articles drawn not only from the history of technology, but also from sociology, political theory, the philosophy of science and of technology, and literature.[13] As such, it pre-sented a "state of the art" of feminist perspectives on technology, ranging from essays on women inventors and engineers to appraisals of the effects of office automation on clerical workers and of household technology on women's domestic labor, from ecological issues and their connections with female values to current issues of technological assessment, reproductive technologies, and feminist concepts of future technology. The introductory essay offered a critique of the masculine bias of literature in the field, showing how feminist perspectives might speak to both the subject matter and the methodological and disciplinary concerns of technology study and research. The closing essay delineated areas for future research, suggesting further ways feminist perspectives might influence and enrich the discipline. The sixfold impact of gender analysis on the study of technology explored below continues and enlarges the discussion thus begun.

EXTENDING THE TECHNOLOGY SPECTRUM: A SIXFOLD IMPACT

Gender analysis in technology started with a focus on women, reflecting the simple fact that they were a neglected subject in the history, philosophy, and sociology of technology. Asking a series of questions, feminist scholars soon found that women had a different relationship to technology and tech-nological development and change by virtue of the fact that, as women, they constituted a particular social category. As it became clear that at the heart of this difference was a set of socially constructed sex roles, behaviors, and activities, the term gender became a conceptual category in pursuing further research. Such an approach fitted well with the view of technology as a social construct that reflects the dominant material forces and cultural values of a society—gender clearly among them. Joining the categories of analysis in this way, gender focus begins to affect other categories in the study of technology. The result has been to extend and transform technology research in the six ways mentioned above: introducing race and class, as well as gender, issues; extending research cross-culturally; enlarging the subject matter; modifying views of technological development and its impact; challenging and changing concepts of technological change; and questioning and broadening the philosophy and values of technology. Selected examples of feminist scholarship will illustrate these impacts.

An important body of research has grown up about the effects on workers of new technologies and technological systems. Computer technology is a recent area of focus. Inquiry extends from workers in the new high-tech industries to different segments of the labor force in the wide variety of workplaces in which automated processes and systems are being introduced. Feminist scholarship adds a critical dimension to exploring the relationship of technology and labor-market segmentation. The initial focus on women provides an opening wedge to examine issues of class, race, and ethnicity, as well as of gender. This occurs because the segmentation and segregation of women workers in a limited number of jobs at the lowest end of the pay, skill, and status scales often reflects women's race and class, with which their gender interacts.

Sally Hacker's research illustrates these interconnections. Studying the effects of automation on women workers in four industries in the United States—insurance, printing and publishing, agribusiness, and telecommunications—Hacker found a number of patterns that often linked gender, race, and class. In agribusiness, among the poorest members of the labor force, often minorities (e.g., migrant, Hispanic), job reduction affected both women and men. But large-scale agrotechnology favored men over women farmers.[14] At the telephone company, over a three-year period automation resulted in a net gain for male workers and corresponding net loss for female workers.[15] As jobs were deskilled in telecommunications and publishing, women moved into new jobs thus created, only to find these jobs eliminated in new rounds of automation. The more sophisticated technical jobs that were created tended to be filled by men. Summing up this research, Hacker wrote, "the flow of workers through occupations during technological change appeared to proceed from white male to minority male to female, then machines."[16]

The electronics industry presents a different labor pattern, relying on a semiskilled workforce rather than on automation and job reduction. Not only is this workforce heavily female, but the workers are also increasingly drawn from recent Spanish-speaking or Asian immigrants to the United States—as in Silicon Valley—or from women living overseas, especially in Southeast Asia, who form a part of the process of "offshore sourcing" in the new internationalized division of labor.[17] Gender inquiry reveals how race and ethnicity as well as class intersect with gender to shape the labor force in the new production system of the electronics industry.[18] These patterns hold similarly for other internationalized, female-dominated industries, notably textile and garment manufacture.[19]

As feminist research thus considers the power-*less* as well as the power-*ful* in exploring the interrelationships of work, workers, and new technologies and technological systems, such research necessarily links industrialized nations and the third world. It therefore extends technological inquiry cross-culturally, the second area of impact of gender analysis. Increasing its attention to the subject of women, technology, and development, this research helps us to readjust our Western lenses to reassess the role of technology transfer and technological innovation for the developing world. Who or what benefits, and how? Solar units for milling grain in Burkina Faso (formerly known as Upper Volta), new rice-processing methods in Indonesia, or training in new farming organization and techniques in many parts of the third world may bring greater "efficiency," but such measures may also eliminate women's income-producing jobs and contribute to greater economic and physical hardships.[20] As such mixed effects of technology transfer cause us to question further Western ideas of technological "progress," we find additional reasons to pursue historically oriented cross-cultural research into traditional technologies. In preindustrial cultures, past and present, traditional technologies are connected to life-sustaining activities: cultivating and processing food, spinning and weaving cloth, building and tending shelters, caring for and nurturing kin. These activities, and their technologies, are predominantly those of women.

Thus, contemporary cross-cultural gender inquiry pulls us back in time to explore further the dimensions of traditional technologies and, in so doing, reminds us of how gender analysis extends the subject matter of technological inquiry, the third area of impact. When unpaid labor is accounted for, women worldwide are responsible for two-thirds of all working hours.[21] Nevertheless, women's activities and their attendant technologies were long neglected in most technological inquiry. Focusing on women's work, therefore, necessarily enlarges the subject matter for technological research. The result is increased attention to a whole range of subjects, including household, reproductive, and office technologies. But the subject categories themselves change and expand, broadening the range in a further way. Medical technology is a case in point. Taking its cue from the scientifically oriented and technically sophisticated medical practices today, research into medical technology has tended to obscure or omit the long history of cures and healing techniques of traditional healers, who have often been women in many cultures.[22] When feminist research reveals that women used moldly bread to treat infections long before the discovery of penicillin or that they knew about ergot for labor pains and digitalis for heart ailments, we add appreciably to our store of knowledge of medical techniques. More important, we expand the range of research in medical technology to include new sources and new subjects. We begin, therefore, to redefine what constitutes the categories and how we approach the categories themselves.

This kind of rethinking also informs the way we look at the social and economic impact of technological developments, the fourth area affected by gender analysis. Even when a determinist framework has not been intended, infrequent references to women and technology have fallen back on a simple cause-effect model, on the basis of insufficient data. For example, the typewriter has often been cited as the cause for women entering clerical work in large numbers starting in the 1880s and even as the cause for the rapid feminization of the clerical labor force that followed. Feminist research, such as the work of Margery Davies, does, indeed, assign the typewriter an important role in the process.[23] As a new technology, it was not yet "gendered": women could become "typewriters" because they were not challenging or replacing men. But technology was not the only contributing factor. Others included the post-Civil War business expansion that proliferated paperwork; the lack of enough men—the traditional clerks—to fill the jobs created; the availability of a substantial pool of native-born, white, literate women for whom such employment would mean an advance, given the limited range of jobs open to them; the tradition that employers, following patriarchal practice, could pay women less than men; and, therefore, the breaking of the taboo and the bending of the ideology that had declared the office an unsuitable milieu for women. In this manner, gender analysis serves to caution against linear cause-effect approaches while promoting a multi-faceted model and suggesting further ways to analyze the interaction of the new technological developments and social phenomena.

Gender analysis brings a similarly cautionary approach to accepted patterns of technological change, the fifth area of impact. When Cowan used quotation marks to enclose *Industrial Revolution* in the title of her article, "The 'Industrial Revolution' in the Home," she alerted us to the ironies involved. Her article challenges fundamental assumptions of that "revolution" when they are applied to mechanization and women's work in the home. The Industrial Revolution, we had all learned, brought centralization of work, specialization, and greater efficiency and productive capacity. Yet the introduction of machinery and new technical systems into the household separated each household and women's work within it still further than previously, creating decentralization, not centralization. The number of machines a woman was to use and jobs she was to perform did not shrink, but proliferated, making her a "Jane of all trades." Further, the isolated woman performing these many jobs only for the benefit of her immediate family was hardly an efficient producer in industrial terms. Exploding another set of myths, research by Cowan and others has revealed the lack of evidence that such "labor-saving" devices increased women's leisure—indeed, there was "more work for mother"—or that technology "freed" middle-class women to enter the paid labor force.[24] Thus, research that began with an inquiry into the relationship of the changes in household technology to women's labor in the home went beyond the original questions posed, as the new empirical data suggested a different set of patterns than traditional theories would have anticipated. Consequently, we are charged with rethinking the nature of technological change and our concepts and theories explaining such change.

When we extend the range of technological inquiry and challenge its theories and assumptions, we begin to touch on fundamental issues of the philosophy of technology, the final area of impact. Technology is a human activity. What can gender analysis tell us about its dimensions? What can we learn about the nature of human interactions with the technology and technological systems that human society has created? Because gender focus raises the quesiton, Who is interacting?, we begin by looking at the language of technological inquiry.

In a representative sampling from *Technology and Culture* and other standard literature, my research has revealed almost exclusive use of male pronouns and indefinite nouns. Generic use is claimed and that terms such as *man* include both sexes. But what are we to make of the fact that specific illustrations of people using technologies—whether simple tools or computers—are almost invariably males?[25] If females and female-associated activities and technologies are omitted from specific reference, can the generic claim stand? Use of male language forms signifies that the abstract human image as well as the specific one in technological inquiry is literally male.

Revealing this literally male character of the human actor in the technology literature leads to a far more significant point. Half of humanity is left out of philosophic speculation: the view of humanity is a truncated one, it is incomplete. If we recall the way gender and sex are confounded so that women's and men's roles and activities are differentiated—and usually hierarchically structured—in every culture, then women's and men's *experiences* will be different. They will experience and approach technology differently, whether it is a birth control device, a computer, or a nuclear missile. Although gender is not the only factor to be considered in reappraising what we mean by the term *human* in the human-technology equation, it is a key factor that has been ignored. Gender perspectives not only call attention to the presently truncated image of that human being, but also seek to fashion a more inclusive human image that considers the experiences as well as the values of both genders.

Because socially constructed gender attributes are attached to each sex, the male image, which is made to stand for all human beings, embodies male gender attributes and perspectives. In the Western study of technology, these are attributes and perspectives of a white, elite male. Western male perspectives on technology are informed by belief in rationality and objectivity (to the exclusion of their complementary opposites), belief in the superiority of mind over body, and, above all, belief in gaining power over nature. Feminist research has illuminated how a hierarchic gender dualism is imbedded in such beliefs and perspectives. Nature, personified as *she,* takes on new significance. Carolyn Merchant has shown, for example, how language and meanings changed as the image of nature was transformed during the Scientific Revolution. Reverence for Mother Nature and sanctions against despoiling her were supplanted by an image of nature as an inert and passive entity that could be penetrated and invaded, her secrets unlocked, her wildness tamed and controlled.[26] With this transformation came greater distancing between mind and body, a separation and alienation of "man" from nature, a drive to gain power over nature—expressed in gendered language and often sexual imagery as well.

A male-female opposition thus attaches to such frequently used metaphors as *man-nature,* or even *man-machine,* as we personify our machines as *she.* The significance of this hierarchic gender dualism for technological inquiry has two aspects. It underscores the dominance of one limiting set of values: technology as conquerer and subduer of nature. Further, gender dualism points to the prevalence of the oppositional model as a philosophical framework: "man" is set against machines, against and over nature. In revealing how deeply gender values are intertwined, gender analysis adds a new dimension to the questioning of this oppositional, power-over model as a framework for philosophic inquiry. Showing that this model and these values are not universal, feminist perspectives seek a concept of human nature that will widen the range of values and theoretical models for the philosophy of technology.

GENDER AND TECHNOLOGY AND BEYOND

When the history of technology emerged as a distinct discipline with the founding of SHOT and the journal *Technology and Culture* in the late 1950s, it linked the disciplines of engineering and history. In seeking a framework in which the interplay of technological innovation and development and of social and cultural phenomena could be successfully pursued, the new discipline creatively broke disciplinary bounds. Sociological and philosophical inquiries into technological phenomena soon also found a home in the history of technology, as, later, did work in other liberal arts disciplines. The disciplinary range of the study of technology was thus further extended. Into this multidisciplinary framework has come gender analysis. As implied in this essay as I have undertaken to illustrate the impact of gender analysis on technology, gender perspectives by their very nature are multi-disciplinary and interdisciplinary. As women's studies has grown and its perspectives have been applied in increasing numbers of disciplines, the findings about gender roles and practices carry over from one field to another. Thus, for example, the psychological and sociological aspects of gender can be useful whether we are examining the impact of gender on literature or on technology. In this manner, gender analysis has also enlarged and extended the disciplinary perspectives of technology. My review of the sixfold impact of gender analysis illustrates this expansion.

Yet, even as it extends the discipline, gender analysis sharpens the focus of technological inquiry. When gender analysis demonstrates that certain groups have been excluded, and that certain questions have not been asked, it reveals the danger of universalizing technological experience and phenomena on the basis of only a selected segment of human experience and human history. Considering women's experiences with technology in the home, it warns us that our models for mechanization and industrialization may be less than universal. It reminds us that depicting "man" against nature as "her" may signify man's experience of nature, but not woman's. Such characterization, too, is less than universal. Gender analysis provides its critique of universals by focusing on the particular, by asking us to question and, if necessary, to redefine our terms and concepts. Through sharpening our focus in this way, gender perspectives push the discipline to add missing dimensions, to change inadequate frameworks, and thus to create a wider vision of technology and the human condition.

NOTES

1. Gayle Rubin, "The Traffic in Women: Notes on the 'Political Economy' of Sex," in *Towards an Anthropology of Women,* ed. Rayna Rapp Reiter (New York: Monthly Review Press, 1975), pp. 157–210.

2. Peter F. Drucker, "Modern Technology and Ancient Jobs," *Technology and Culture* 4 (Summer 1963): 277–81; Alison Ravetz, "Modern Technology and an Ancient Occupation: Housework in Present Day Society," *Technology and Culture* 6 (Spring 1965): 256–60.

3. Ruth Schwartz Cowan, "A Case Study of Technological and Social Change: The Washing Machine and the Working Wife," paper presented at Berkshire Conference of Women Historians, Douglass College, Rutgers University, March 1973, and published in *Clio's Consciousness Raised: New Perspectives on the History of Women,* ed. Mary Hartman and Lois W. Banner (New York: Harper & Row, 1974), pp. 245–53; Ruth Schwartz Cowan, "Household Technology and Women: A Case Study in Technological Determinism," paper presented at sixteenth annual meeting, Society for the History of Technology, San Francisco, December 1973.

4. Ruth Schwartz Cowan, "The 'Industrial Revolution' in the Home: Household Technology and Social Change in the 20th Century," *Technology and Culture* 17 (January 1976): 1–23.

5. Ruth Schwartz Cowan, *More Work for Mother: The Ironies of Household Technology from the Open Hearth to the Microwave* (New York: Basic Books, 1983). See also the important contributions of Joann Vanek, "Time Spent in Housework," *Scientific American* 231 (November 1974): 116–20; and Susan Strasser, *Never Done: A History of American Housework* (New York: Pantheon, 1982).

6. "Women in Technological History," *Technology and Culture* 18 (July 1977): 496–97.

7. Martha Moore Trescott, ed., *Dynamos and Virgins Revisited: Women and Technological Change in History* (Metuchen, N.J.: Scarecrow, 1979).

8. Judith A. McGaw, "Women and the History of American Technology," *Signs: Journal of Women in Culture and Society* 7 (Summer 1982): 798–828.

9. "Women in Technological History."

10. See discussion in Joan Rothschild, "Introduction," *Machina Ex Dea: Feminist Perspectives on Technology*, ed. Joan Rothschild (New York and Oxford: Pergamon, 1983), p. xv.

11. Trescott, *Dynamos and Virgins Revisited;* see especially Ruth Schwartz Cowan, "From Virginia Dare to Virginia Slims: Women and Technology in American Life," pp. 30–44. This article, appearing originally in *Technology and Culture* 20 (January 1979): 51–63, was adapted from Cowan's earlier presentation at the bicentennial meeting [eighteenth annual] of the Society for the History of Technology, Washington, D.C., October 1975.

12. McGaw, "Women and the History of American Technology."

13. Rothschild, *Machina Ex Dea.*

14. Sally Hacker, "Farming out the Home: Women and Agribusiness," *The Second Wave* 5 (Spring/Summer 1977): 38–49.

15. Sally Hacker, "Sex Stratification, Technology and Organizational Change: A Longitudinal Case Study of AT&T," *Social Problems* 26 (June 1979): 539–57.

16. Sally L. Hacker, "The Culture of Engineering: Woman, Workplace and Machine," *Women's Studies International Quarterly* 4 (1981): 341–53. See the opening pages of this article for further summary of Hacker's research on the effects of technological change on the workforce in these industries.

17. In "offshore sourcing," the components are shipped overseas for assembly, the finished products returned to the home country for sale. See discussion on the electronics industry and the labor force in Robert T. Snow, "The New International Division of Labor and the U.S. Work Force: The Case of the Electronics Industry," in *Women, Men, and the International Division of Labor*, ed. June Nash and María Patricia Fernández-Kelly (Albany, N.Y.: State University of New York Press, 1983), pp. 39–69. See also the articles in part 4 of the Nash and Fernández-Kelly volume.

18. See the perceptive theoretical analysis of Susan S. Green, "Silicon Valley's Women Workers: A Theoretical Analysis of Sex-Segregation in the Electronics Industry Labor Market," in Nash and Fernández-Kelly, *Women, Men, and the International Division of Labor*, pp. 273–331.

19. See Wendy Chapkis and Cynthia Enloe, eds., *Of Common Cloth: Women in the Global Textile Industry* (Amsterdam: Transnational Institute; Washington, D.C.: Institute for Policy Studies; London: Pluto Press, 1983).

20. See these and other examples in Roslyn Dauber and Melinda L. Cain, eds., *Women and Technological Change in Developing Countries* (Boulder, Colo.: Westview Press, 1981).

21. As reported in Development Issue Paper No. 12, United National Development Program, United Nations: Women represent one-half the global population and one-third of the labor force, but are responsible for two-thirds of all working hours, receive only one-tenth of world income, and own less than one percent of world property.

22. See, e.g., Autumn Stanley, "Women Hold Up Two-Thirds of the Sky: Notes for a Revised History of Technology," in Rothschild, *Machina Ex Dea*, pp. 3–22.

23. Margery W. Davies, *Woman's Place Is at the Typewriter: Office Work and Office Workers, 1870–1930* (Philadelphia: Temple University Press, 1982).

24. Cowan, *More Work for Mother;* Vanek, "Time Spent in Housework"; Rothschild, "Technology, Housework, and Women's Liberation: A Theoretical Analysis," in Rothschild, *Machina Ex Dea*, pp. 79–93.

25. Rothschild, "Introduction," *Machina Ex Dea.* See further discussion in Rothschild, *Teaching Technology from A Feminist Perspective: A Practical Guide* (New York and Oxford: Pergamon, 1988), chap. 6.

26. Carolyn Merchant, *The Death of Nature: Women, Ecology, and the Scientific Revolution* (New York: Harper & Row, 1980).

Its Own Reward: Three Decades of Teaching and Scholarship in the History of Technology

DARWIN H. STAPLETON
with
LIZ PALEY

At least since the founding of the Society for the History of Technology (SHOT) in 1958, the study of the history of technology has thrived, primarily in academia. Melvin Kranzberg, the primary force behind SHOT's establishment, was then at Case Institute of Technology (Cleveland) where faculty and administrators had from about 1955 developed and nurtured an undergraduate humanities curriculum that emphasized the role of technology (and science) in world history. (In 1961 Case established a graduate degree in the history of science and technology, usually cited as the first graduate degree in the Western world that formally incorporated the history of technology. Kranzberg was joined in founding SHOT primarily by others drawn from academic ranks and engaged in undergraduate education, including Car Condit, Morrell Heald, Hugo Meier, and John B. Rae. The seedbed for the modern discipline of the history of technology in the United States was thus undergraduate education, although the membership of SHOT soon included engineers, scientists, and museum professionals, as well as educators.[1]

The self-conscious establishment of an academic discipline is often associated with increasing specialization, with the creation of esoteric language with conferences and graduate seminars, and with the founding of journals and societies intended to justify spinning off another fragment of human knowledge from what remains accessible to the literate public. More to the point, a new discipline is usually thought to distance itself rapidly from undergraduate education, because it becomes too arcane to address in survey courses, and its literature is often obscure even to the brightest undergraduates.

Moreover, both academics and outsiders agree that there is a Peter Principle at work inside the ivy walls.[2] One becomes a professor in large part because one is good at studying, writing exams, and grinding through the research and writing of a dissertation, while the opportunities for acquiring teaching experience are regarded as distinctly secondary to the real purposes of graduate school. Once ensconced in a groove in the academic track, opportunities for advancement are based primarily on publishing books and articles, serving the institution as required, and teaching effectively, in that order.[3] Research and publication lead to promotion and peer approval; teaching must be its own reward.

That publication is perceived as the key to academic survival is evidenced by the poignant rationale for a 1985 conference session regarding "Research in STS [science, technology and society] Studies." A participant wrote, "It was the central proposition of [that] Conference session that if STS studies is to achieve a secure position alongside the traditional academic disciplines, it needs to establish research credentials in addition to its already well-deserved reputation for innovative curriculum design."[4]

My purpose in this essay is to explore the extent to which the history of technology in the United States has been able to retain its roots in undergraduate education while developing a body of scholarship, first by examining how often the products of academic scholarship in the field have been useful in the classroom, and second by considering ways in which the field as it exists today (1988) is engaged in developing and promoting tools for teaching. This examination suggests that American historians of technology have created much scholarship that is useful in the classroom, and that they have retained a lively interest in the materials that can enrich the teaching of the subject.

The most comprehensive examination of the way the history of technology has been taught in the United States is Svante Lindqvist's *The Teaching of History of Technology in USA—A Critical Survey in 1978*. Compiled after Lindqvist's tour of centers for the history of technology and technology studies in the United States (and aimed at providing those contemplating creating new centers abroad with a guide to what was current practice), the survey is a personal and detailed look at seventeen programs. Several of Lindqvist's conclusions regarding the situation in 1978 are relevant here: that the field had no textbooks regarded as standard, with the possible exception of Melvin Kranzberg and Carroll W. Pursell, Jr.'s *Technology in Western Civilization;* that readings for courses were taken from a wide variety of publications, only a portion of which might be regarded as historical; that audiovisual materials were used extensively; that few courses included visits to museums or historic sites; and that the history of technology in American universities tended to be taught as *the history of technology in America.*[5]

The greatest contribution to undergraduate history curricula that a scholar can make is to provide a textbook for the teachers and students who have not spent sufficient time in Clio's thrall to integrate all aspects of the subject. This need is perhaps felt most strongly by academics who enter the history of technology via engineering (as Lindqvist himself did), a field that has been taught by the textbook method for generations. Because many programs in the history of technology are closely related to engineering schools, a large portion of undergraduates also expect to find a textbook matched with a syllabus. To encounter a course that does not begin by rigorously defining its field and the underlying principles, and then proceed by elucidating those principles, can be mystifying to the prospective engineer.[6]

Although the availability of one or more standard textbooks for the history of technology would help alleviate that distress, no such volume presently exists (although several are reported to be in progress at this writing). Interestingly enough, the coalescence of the field nearly thirty years ago coincided with the publication of the first of several surveys. The most monumental was *A History of Technology,* edited by Charles Singer and colleagues, with chapters written by a large group of specialists. It was conceived as a work for the student who was being "professionally trained in science or technology," although, as Robert Woodbury pointed out, the five volumes of about 750 pages each are hardly designed for the typical technical education curriculum.[7] The thorough reviews of *A History of Technology* that Melvin Kranzberg commissioned for an early issue of *Technology and Culture* demonstrated that, while the work was a monumental compendium, it was also a codification of error and a tacit admission of inadequate underlying scholarship.[8]

Kranzberg's editorship of *Technology and Culture* influenced his subsequent plan for the two-volume text that he coedited with Carroll W. Pursell, Jr., *Technology in Western Civilization* (1967).[9] Arising in an American context (whereas *A History of Technology* was British), this survey covered all human history to 1900 in the first volume, and the twentieth century in the second, centering on American contributions to industrialization and mechanization. The topics addressed were undoubtedly a reflection of the syllabi of the history of technology courses Kranzberg and Pursell taught at Case Institute of Technology, but they also reflected the various specialties in the history of technology then being cultivated by American academics.[10]

The Kranzberg-Pursell volumes were far better oriented to the classroom than Singer's and immediately were adopted by many teachers. By 1979 Kranzberg-Pursell was the most used history of technology book (according to Lindqvist's survey), and, although there were many competitors, even in 1986 it continued to be among the leading works assigned in history of technology courses.[11]

It did have competitors, however. As a group, these were largely written by Americans about the history of American technology. Most notable was *Technology in America,* edited by Carroll W. Pursell, Jr. First published in 1977 as a Voice of America book for readers whose first language was not English, it was re-published in 1979 in virtually the same form by MIT Press for the domestic market. Almost all of the authors in *Technology in America* could be identified as professors of the history of technology, in contrast to many of the essayists in *Technology in Western Civilization* a decade earlier, who had to be drawn from ancillary disciplines.[12]

Indeed, an increasing portion of the books that have become available for history of technology classes appear to have been written by historians of technology. Works by the likes of Lewis Mumford (identified with architectural history, urban history, and social criticism), Carlo Cipolla (economic history), and V. Gordon Childe (archaeology) are being replaced by those of a new generation of American historians of technology such as Merritt Roe Smith, Ruth Schwartz Cowan, and David A. Hounshell.[13] At the same time there are classics written or compiled by historians of technology that remain favorites, including Edwin T. Layton, Jr.'s *Technology and Social Change in America,* Lynn White's *Medieval Technology and Social Change,* and D. S. L. Cardwell's *Turning Points in Western Technology.*[14] Still, a good deal of worthwhile and provocative material appropriate to the history of technology curriculum is being published by scholars not clearly identifiable with the field. Books in American studies or American civilization account for much of it—for example, John Kasson's *Civilizing the Machine,* James Flink's *The Car Culture,* and Daniel Boorstin's trilogy, *The Americans.*[15]

The continued influx of good history of technology written by scholars not committed to the field has disturbed some insiders.[16] But I believe that the history of technology classrooms in the United States probably could benefit from the use of more publications written by scholars outside the discipline, especially scholars from fields that are largely nonmodern in orientation. As Lindqvist noted in the late 1970s, the courses taught here have tended to focus on American technology, presumably because students evince more interest in Model Ts than ziggurats. Certainly, looking at our own history brings important recent developments into the classroom (e.g., nuclear energy, pollution, computerization, medical technology). On the other hand, the Americanization of the history of technology runs counter to the fundamental premise that the technology of the modern era is built on the contributions of various cultures throughout the knowable past.[17]

In any case, recent scholars in the history of technology have not provided students a single, comprehensive textbook, either for American technology or Western technology, let alone the course of technology in history. Derek Price wrote some time ago that hopes for such a book were "barren" considering the failure of the history of science to produce a general text, and at this juncture he appears to have been correct.[18] On reflection, the absence of a textbook may be beneficial. As a consequence, teachers of the history of technology must be constantly aware of new materials (including old books reissued as paperbacks), searching the annual bibliographies in *Technology and Culture,* checklists and reviews in newsletters, tables of contents of *Scientific American,* and national periodicals and newspapers.[19] Perhaps this is one reason why SHOT (if it may be taken as representative of teaching historians of technology) appears to be so open to new faces, new ideas, and new approaches to the field.[20]

Clearly, journal articles and chapters excised from books are a staple of most history of technology courses. Where do these selections come from? In contradiction to the thesis that as fields develop they become increasingly arcane and inaccessible to the intelligent layperson, *Technology and Culture,* the initial American journal in the field and still the leading journal, remains the most-used scholarly journal for undergraduate courses.[21] The journal's early and continued dedication to the publication of articles that raise major questions or are fundamentally interdisciplinary probably contributes much to its value in the classroom. Older articles from *Technology and Culture* by Heilbroner, Cowan, and Layton are regularly listed in syllabi, and more recent pieces by Mazuzan on nuclear power and Vincenti on the nature of engineering knowledge could quite easily become classroom classics.[22]

Other scholarly journals regularly mined for the classroom include *Science, Isis,* and *Business History Review,* each of which has been receptive to articles with a major focus on the history of technology. For example, *Science* published Eugene S. Ferguson's "The Mind's Eye: Nonverbal Thought in Technology," which is extremely useful in bringing to students' attention not only an important aspect of cognition in technical activities, but also the difference between science and technology, and the importance of using visual materials to study technology.[23]

It is precisely the matter of seeing the technology as well as reading about it that leads many professors to rely heavily on articles in *Scientific American.* Numerous specialists in the history of technology who have published more academic analyses elsewhere have written stimulating and incisive summaries of their research for *Scientific American.* That magazine has insisted on heavy use of illustrations, including historical prints as well as modern diagrams, creating a feast of visual information that few scholarly journals can approach. Historians of technology who do not subscribe to *Scientific American* must make an annual or semiannual pilgrimage to their libraries to search for new gems for their syllabi.[24]

Journals such as *American Quarterly, American Historical Review, Journal of American History, Journal of Economic History, Technology Review,* and *Technikgeschichte* (which publishes some articles in English) also have their adherents. Clearly, good articles for the classroom come from a variety of sources. I find most useful articles that are clearly written, that provide sufficient historical evidence so that the student has enough data to reach conclusions different from the author's, and that put the particular subject in the context of larger historical patterns or issues.[25]

Although good books and articles for the classroom abound, there are always situations for which nothing published will do or that require teaching tools (such as visual aids) that are normally not provided by the marketplace. In those situations most professors create their own materials. The "course packet" or "course reader" compiled by the professor may include some combination of essays and notes written specifically for the course; articles and excerpts from other publications; illustrations, charts, and tables; the usual handouts, such as syllabus, bibliography, and assignment sheets; and even examination questions. The advent of cheap, high-quality photocopying, and particularly of quick-copy stores that provide bound packets to students, has made the course packet easy to provide, however time-consuming it may be to assemble. The recent development of "desk-top publishing" programs for word processors may create additional possibilities for this sort of curriculum development. Moreover, at least one company has sought out course packets to consider for its list of publications.[26]

According to Lindqvist's survey, the area of course creativity about which historians of technology are most proud are their slide collections. Some university departments create and maintain extensive collections, but normally slides are made to complement individual lecture notes and discussions. Even so, there are some constants, such as the need for line drawings, or other simple but strong images; illustrations of technology created in the period in question; and at least one good picture of a Newcomen engine![27]

Even though the Aristotelian method of instruction most common in universities is verbal and unidirectional, visual images are nonverbal and at least raise the possibility that the classroom experience will be participatory. Pictures convey the appearance of a building or the operation of a machine as no sequence of words can, and for many students (perhaps especially those with technical or artistic training), pictures can be used to draw their eyes into, around, and through a process or device. I have always found it effective with moderate-sized classes (ten to twenty-five students) to display a slide on the screen and to point out a number of elements that integrate the slide with the previous discussion or lecture, and then ask for questions or comments. Illustrations from *De Re Metallica* or da Vinci's notebooks are so rich in information that entire sessions can be constructed from the examination of ten to twenty slides.[28]

While slides are extremely useful, there is nothing like viewing "a piece of the true Cross."[29] Everyone has nearby some technological site worth visiting, whether it is a pioneer log cabin, the university's boiler room, a museum of antique automobiles, or a nuclear power plant. Yet few professors take as much time to utilize sites as they do slides, perhaps because their training and experience are not object-oriented or the sites do not always teach the lessons they want to have learned, and certainly because arranging extramural activities is time consuming. Nonetheless, site visits are rewarding. When I taught at Case Western Reserve University in Cleveland, the following "artifacts" were within walking distance of campus: prehistoric tools (Cleveland Museum of Natural History); Mesopotamian, Egyptian, and

Greco-Roman crafts (Cleveland Museum of Art); a replica of a Gothic chapel (Case Western Reserve University); early automobiles (Western Reserve Historical Society); and a display of twentieth-century medical technology (Dittrick Museum of Historical Medicine). In several instances the curators of the collections were available to give authoritative introductions to, or tours of, the site and objects. Students were required to visit the sites and generally responded enthusiastically.[30]

The annual meeting of SHOT may also have a significant impact on the organization of courses. Although meetings had few sessions during SHOT's first dozen or so years, and (so far as I can tell from published accounts) none devoted to teaching, several sessions in the years after 1973 have considered aspects of the history of technology and technology studies curricula. Most of them have been upbeat, describing the origins and rise of new programs and considering the new opportunities raised by such elements of the programs as the interdisciplinary nature of the faculty.[31] The creation of the Technology Studies and Education interest group within SHOT in 1977–78 has resulted in regular sessions devoted to teaching.[32]

Yet relatively little attention is given at the meetings to curriculum development and classroom practice. One suspects that there is a general reluctance to preach to those who already feel themselves experienced teachers. Moreover, criticism of teaching is rare among faculty. In contrast to well-developed traditions of scholarly criticism, including peer review of papers, published reviews of books, and formal commentary on papers delivered at professional meetings, it is unusual for professors to have their teaching or curricula critiqued by their departmental peers (often their friends), let alone their disciplinary peers.

Perhaps it comes down to the difference in audiences for the activities of teaching and scholarship. The audience for teaching is the students, and effective teaching is normally rewarded by sustained (even increased) enrollments. In some universities students even rate courses (and professors) and publish the findings. Having been critiqued by undergraduates, it may be too much to suffer the potential reproof of colleagues.

It is also worth recognizing that many historians of technology, perhaps most, teach their discipline as its solitary representative on the institution's faculty. Neither day-to-day monitoring nor annual departmental review of curriculum may be appropriate or helpful. The real sharing of ideas about curricula, and one that undoubtedly invites self-criticism and stimulus for improvement, comes from newsletters which publish syllabi, reading lists, bibliographies, and notes about teaching.

The most used newsletter in the history of technology and technology studies is *Science, Technology & Society,* published at Lehigh University since 1977.[33] The first issue noted that the editor intended to publish short articles on the theoretical and speculative aspects of curriculum development, in-depth course descriptions, reviews of texts and audio-visual aids, and current bibliography (annotated). In addition, we would welcome articles on successful techniques for such tasks as instituting and evaluating a course or program, arousing faculty and student interest, overcoming administrative reluctance, obtaining visibility on campus, running a lecture or film series, or editing a newsletter.[34]

One may see that in this statement, like the sessions of SHOT, there is at least as much interest in sharing the vicissitudes of organization and bureaucracy as in sharing the results of teaching. Nonetheless, the *STS Newsletter* has been successful in treating a variety of topics in teaching. One of the most impressive has been the regular publication of syllabi, a major stimulus to and source for the Technology Studies and Education group's compilation of model syllabi as *The Machine in the University* (1983). The demand for that collection has been substantial, leading to publication of a revised edition only four years later.[35]

Other newsletters that have been useful to historians of technology include *The Weaver,* which publishes discursive essays on subjects in modern technology that should be examined in the classroom; *The Charles Babbage Institute Newsletter* (from the Center for the History of Information Processing), which brings readers up-do-date on the newest publications in the history of computing; and *History of Science: News & Views,* which aims to establish a network of scholars in the history of American science (and considers broadly defined history of technology as within that pale).[36]

Clearly, formal scholarly processes such as publications and national meetings do not provide the forum for curricular development that informal processes such as newsletters and personal contact do. In a pattern parallel to that of the "mirror-image parity" that Edwin Layton proposed for the value systems of science and technology in late nineteenth-century America, scholarship and teaching are treated differently in the history of technology community in the United States.[37] Scholarship is subjected to peer review and published in books and journals for the world to read; teaching is personally developed, aimed at a student audience that may have little previous knowledge of the subject (yet may be highly critical), and shared with others only through informal publications limited to a mailing list of "insiders."

This inversion of values makes the connections between teaching and scholarship modest, even when scholar and professor are one. On the one hand, the body of publications produced in the three decades since the founding of SHOT has contained many titles that are valuable in the classroom, and the leading journal in the field, *Technology and Culture,* has retained the breadth that has allowed many of its articles to be accessible and stimulating to undergraduates. On the other hand, historians of technology have been no more effective than other academicians at turning the traditional disciplinary bureaucracy and apparatus to the task of creating, nurturing, and sustaining good teaching. Even the SHOT sessions and the curriculum newsletters apparently devoted to teaching have a tendency to be diverted from activities of the classroom to the politics of undergraduate education. As historians of technology we are effective at identifying and celebrating scholarship; but our pedagogy receives far less attention. Good teaching is, indeed, largely its own reward.

NOTES

1. *Technology and Culture* 1 (Winter 1959–60): 106–7; T. K. Glennan, "To all Case Men," 16 December 1952, box 14, Glennan Office Files, Case Western Reserve University Archives, Cleveland, Ohio; Morrell Heald, interview by author, 26 June 1986; Melvin Kranzberg, interview by author, 17 October 1986.

2. J. Lawrence Peter, *The Peter Principle: Why Things Always Go Wrong* (New York: Morrow, 1969).

3. Svante Lindqvist, *The Teaching of History of Technology in USA—A Critical Survey in 1978*. Stockholm Papers in History and Philosophy of Technology (Stockholm: Royal Institute of Technology Library, 1981), p. 26.

4. Stephen H. Cutcliffe, "Research in STS Studies Theme Issue," *Science Technology & Society Newsletter* 52 (February 1986): 1. The session was at the first annual meeting of the National Technology Literacy Conference.

5. Lindqvist, *The Teaching of the History of Technology in USA*, pp. 43–46, 51–70, 88–90, and passim.

6. In ten years (1976–86) of teaching the history of technology at Case Western Reserve University, in classes largely made up of engineering majors, I often heard expressions of bewilderment regarding what the study of history claimed to be, the nature of historical evidence, and the possibility of historical proof. As I reflect on my response to their confusion, I realize that after the first two years, I compensated by distributing and discussing definitions of the history of science and technology, and by giving a lecture, "Principles and Myths of Technology." By presenting the definitions and lecture as a personal viewpoint, but based on reading and research, I legitimized the student's need to make sense of what they perceived as a "fuzzy" discipline, while arguing that there were some concepts that had proved useful to many historians.

7. Charles Singer, "How 'A History of Technology' Came Into Being," *Technology and Culture* 1 (Fall 1960): 306; Robert S. Woodbury, "The Scholarly Future of the History of Technology," ibid., 346.

8. *Technology and Culture* 1 (Fall 1960): 299–414.

9. Melvin Kranzberg and Carroll W. Pursell, Jr., eds., *Technology in Western Civilization*, 2 vols. (New York: Oxford University Press, 1967); Melvin Kranzberg, interview by author, 17 October 1986.

10. *Catalogue*, Case Institute of Technology, 1963–67.

11. Lindqvist, *The Teaching of History of Technology in USA*, pp. 52–55. My assessment of textbook usage in 1986 comes from the results of a survey (hereafter, "survey 1986") mailed out to approximately eighty persons on the *Technology and Culture* mailing list, published in the January 1985 issue, whom I recognized as

teacher-scholars in the history of technology. Forty-five survey forms were returned, of which six contained no responses, the addressees declining to answer for various reasons. A Case Western Reserve doctoral candidate, Liz Paley, compiled the responses (as well as contributing to the essay). In response to the request to "List up to five books in the history of technology which you have found most valuable to assign as course reading, at either the undergraduate or graduate level," there were 87 books listed, only 32 more than once. Kranzberg and Pursell was named eight times, with only four other books mentioned more often; another book was also mentioned eight times.

12. Survey 1986; Carroll W. Pursell, Jr., ed., *Technology in America: A History of Individuals and Ideas* (Cambridge, Mass.: MIT Press, 1981); Robert P. Multhauf, "Some Observations on the State of the History of Technology," *Technology and Culture* 15 (January 1974): 7; Kranzberg and Pursell, *Technology in Western Civilization*, 1:775–79, 2:741–45.

13. Books normally "become available" for the classroom when they are issued in paperback. Few clothcover or hardcover books are priced appropriately for undergraduate purchase; duplicating entire books, or even chapters, for classroom distribution is severely restricted by copyright law.

14. Lewis Mumford, *Technics and Civilization* (New York: Harcourt, Brace & World, 1934); Carlo Cipolla, *Guns, Sails, and Empires: Technological Innovation and the Early Phase of European Expansion, 1400–1700* (New York: Minerva, 1965); V. Gordon Childe, *Man Makes Himself* (New York: Mentor, 1951); Lindqvist, *The Teaching of the History of Technology in USA*, pp. 54–55; survey 1986; Merritt Roe Smith, *Harpers Ferry Armory and the New Technology: The Challenge of Change* (Ithaca, N.Y.: Cornell University Press, 1977); Ruth Schwartz Cowan, *More Work for Mother: The Ironies of Housework from the Open Hearth to Microwave* (New York: Basic Books, 1983); David A. Hounshell, *From the American System to Mass Production 1800–1932: The Development of Manufacturing Technology in the United States* (Baltimore: Johns Hopkins University Press, 1984); Edwin T. Layton, Jr., ed., *Technology and Social Change in America* (New York: Harper and Row, 1973); Lynn White, jr., *Medieval Technology and Social Change* (New York: Oxford University Press, 1962); D. S. L. Cardwell, *Turning Points in Western Technology* (New York: Science History Press, 1972).

15. John Kasson, *Civilizing the Machine: Technology and Republican Values in America, 1776–1900* (New York: Grossman, 1976); James J. Flink, *The Car Culture* (Cambridge, Mass.: MIT Press, 1975); Daniel Boorstin, *The Americans*, 3 vols. (New York: Vintage Books, 1958–73); survey 1986.

16. David A. Hounshell, "On the Discipline of the History of American Technology," *Journal of American History* 67 (March 1981): 854–65. See also the letters that followed: Darwin H. Stapleton and David Hounshell, "The Discipline of the History of American Technology: An Exchange," *Journal of American History* 68 (March 1982): 897–902.

17. Lindqvist, *The Teaching of History of Technology in USA*, pp. 43–48.

18. Derek de Solla Price, "On the Historiographic Revolution in the History of Technology: Commentary on the Papers of Multhauf, Ferguson, and Layton," *Technology and Culture* 15 (January 1974): 44; cf. Lindqvist, *The Teaching of History of Technology in USA*, p. 58.

19. Survey 1986; Stephen Cutcliffe, "Special STS Survey Issue," *Science, Technology & Society Newsletter* 25 (September 1981): 1–2. Cutcliffe commented: "It is fairly clear from the responses [to a survey of science, technology and society courses] that there is no single text or texts that a majority of instructors find particularly useful, within a group of similar courses, e.g., the history of technology or general technology and society type courses."

20. Others may disagree with this assessment, but it stems from more than fifteen years of observing SHOT meetings and is, I believe, in part the result of Kranzberg's philosophy that the discipline can be strengthened by input from other disciplines.

21. Survey 1986. See also *The Machine in the University: Sample Course Syllabi for the History of Technology and Technology Studies*, ed. and comp. Stephen H. Cutcliffe and TS&E Committee (Bethlehem, Penn.: STS Program, Lehigh University, 1983); Lindqvist, *The Teaching of the History of Technology in USA*, pp. 58–59.

22. Robert Heibroner, "Do Machines Make History?" *Technology and Culture* 8 (July 1967): 335–45; Ruth Schwartz Cowan, "The 'Industrial Revolution' in the Home: Household Technology and Social Change in the 20th Century," ibid. 17 (January 1976): 1–23; Edwin T. Layton, Jr., "Mirror-Image Twins: The Communities of Science and Technology in 19th Century America," ibid. 12 (October 1971): 562–80; George T. Mazuzan, "Atomic Power Safety: The Case of the Power Reactor Development Company Fast Breeder, 1955–1956," ibid. 23 (July 1982): 341–71; George T. Mazuzan, "'Very Risky Business': A Power Reactor for New York City," ibid. 27 (April 1986): 262–84; Walter G. Vincenti, "Technological Knowledge Without Science: The Innovation of Flush Riveting in American Airplanes, ca. 1930–ca. 1950," ibid. 25 (July 1984): 540–76; Walter G. Vincenti, "Control-Volume Analysis: A Difference in Thinking Between Engineering and Physics," ibid. 23 (April 1982): 145–74.

23. Eugene S. Ferguson, "The Mind's Eye: Nonverbal Thought in Technology," *Science* 197 (26 August 1977): 827–36.

24. An anthology of classic *Scientific American* articles dealing with technology was published as *Scientific Technology and Social Change: Readings from Scientific American* (San Francisco: Freeman, 1974). More recent articles include: Robert Mark and William W. Clark, "Gothic Structural Experimentation," *Scientific American* 251 (November 1984): 176–84; Terry S. Reynolds, "Medieval Roots of the Industrial Revolution," ibid. 251 (July 1984): 122–30; Frederick J. Hooven, "The Wright Brothers' Flight-Control System," ibid. 239 (November 1978): 167–82; Werner Soedel and Vernard Foley, "Ancient Catapults," ibid. 240 (March 1979): 150–60; Vernard Foley and Werner Soedel, "Ancient Oared Warships," ibid. 244 (April 1981): 148–63.

25. Survey 1986. Examples of articles that fit my criteria include Lynn White, jr., "The Historical Roots of our Ecologic Crisis," *Science* 155 (10 March 1967): 1203–7; James J. Flink, "The Three Stages of American Automobile Consciousness," *American Quarterly* 24 (October 1972): 451–73; William TeBrake, "Air Pollution and Fuel Crises in Preindustrial London, 1250–1650," *Technology and Culture* 16 (July 1975): 337–59.

26. See, e.g., John Staudenmaier's outline of his packet for "America's Technological Style," in *The Machine in the University*, p. 99. For desktop publishing, see Peter H. Lewis, "Publishing Newsletters by Computer," *New York Times*, 2 September 1986. The company I have in mind is Kinko's Publishing.

27. Lindqvist, *The Teaching of History of Technology in USA*, pp. 60–62.

28. Georgius Agricola, *De Re Metallica*, trans. Herbert Clark Hoover and Lou Henry Hoover (New York: Dover, 1950); Jean Paul Richter, ed., *The Notebooks of Leonardo da Vinci, Compiled and Edited from the Original Manuscripts*, 2 vols. (New York: Dover, 1970); Ladislao Reti, ed., *The Unknown Leonardo*. (New York: McGraw-Hill, 1974).

29. I draw my expression from Brooke Hindle, "How Much is a Piece of the True Cross Worth?" in *Material Culture and the Study of American Life*, ed. Ian M. G. Quimby (New York: W. W. Norton, 1978), pp. 5–20.

30. Darwin H. Stapleton, "The City as Artifact: Utilizing Walking Tours and Museums in Teaching the History of American Technology," *Science, Technology & Society Newsletter* 34 (February 1983): 1–4; survey 1986; Lindqvist, *The Teaching of History of Technology in USA*, pp. 68–70.

31. E.g., annual meeting reports in *Technology and Culture* 15 (July 1974): 453–56; 18 (July 1977): 488–90; 21 (July 1980): 458–60; 24 (July 1983): 490–91; 27 (July 1986): 584–85.

32. Stephen Cutcliffe to author, 24 September 1986; Raymond Merritt to Eugene S. Ferguson, 15 December 1977; Raymond Merritt, memorandum to SHOT Technological Studies and Education Committee, 20 November 1978. Copies in author's possession.

33. Survey 1986.

34. *Humanities Perspectives on Technology: Curriculum Newsletter of the Lehigh University HPT Program* 1 (August 1977): 2. (This was the title of the *Newsletter* until mid-1979.)

35. *The Machine in the University: Sample Course Syllabi for the History of Technology and Technology Studies,* ed. and comp. Terry S. Reynolds and TS&E Committee. 2d rev. ed. (Houghton, Mich.: STS Program, Michigan Technological University; Bethlehem, Pa.: STS Program, Lehigh University, 1987).

36. *The Weaver of Information and Perspectives on Technological Literacy* (1982–); *Charles Babbage Institute Newsletter* (1978–); *History of Science in America: News and Views* (1980–).

37. Layton, "Mirror-Image Twins," p. 576.

Rethinking the History of "American Technology"

DAVID A. HOUNSHELL

For a generation now—the "Kranzberg generation"—scholars in the United States have been piecing together a history of "American technology." There have been some remarkable achievements in this scholarship. One need only peruse the pages of *Technology and Culture* to gain some sense of the strength and vitality of the field. But success notwithstanding, there is a gnawing problem that emerges not only from scholarship in the history of technology in America itself but also from two decades of publishing in general American history. The problem is both methodological and ideological. The bulk of scholarship in the history of technology in the United States has been aimed either explicitly or implicitly at developing a concept of the uniqueness of "American technology." This scholarship has been highly nationalistic. Although historians have acknowledged the importance of the transfer of technology from Europe (chiefly from England) to the United States in the antebellum period,[1] they have nevertheless continued to celebrate Yankee ingenuity while suggesting implicitly that new technology in the United States after the Civil War was largely homegrown, democratic, and superior. In this respect, scholarship on technology in America is among the last vestiges of the Progressive historical tradition that began with Frederick Jackson Turner's classic 1893 paper on the frontier in American history.[2]

Progressive history and the consensus history that built upon it during the 1950s and early 1960s have given way to richer schools of historical writing. In political history, Bernard Bailyn, Gordon Wood, and Joyce Appleby, among others, have opened our eyes to the great continuity in political ideas and ideals between England and her American colonies.[3] The new American nation, to quote Appleby, "was not born free, rich, and modern."[4] Rather, it was based on European ideas of class, propriety, and politics. Rhys Isaac's *Transformation of Virginia, 1740–1790* moved beyond the subject of political ideology to describe the way in which Virginians aspired to emulate the English gentry in virtually every aspect of their lives, including their material surroundings.[5] These works underscore the need to look at American history in terms of its relationship with European artifacts, ideas, and traditions.

Such thinking has scarcely penetrated the work of historians of technology, largely because of nationalist sentiments.

Much of the nationalism in our scholarship stems from the construction of our general histories around the 1851 London Crystal Palace Exhibition. The new exhibition at the Smithsonian Institution's National Museum of American History, "Engines of Change," and the companion volume by Brooke Hindle and Steven Lubar, *Engines of Change,* are only the most recent examples of this tendency.[6] According to the standard treatment of the Crystal Palace Exhibition, the United States was highly dependent on Great Britain for its technology up to the mid-nineteenth century. But, as evidenced by the American displays at the Crystal Palace by 1851, the United States had established an independent technical tradition and was ready to assume a position of world leadership. To quote Hindle and Lubar's most recent formulation of this well-worked premise, "The United States had begun to surpass its mentor."[7] Such analysis is hardly different in tone from the highly nationalistic contemporary work, *American Superiority at the World's Fair.*[8] Yet it has become a dominant theme in our scholarship. Nathan Rosenberg's standard *Technology and American Economic Growth* is loosely structured around the London Crystal Palace Exhibition and New York's slavish but second-rate 1853 imitation of the first world's fair. Before the London show, the United States was a net borrower of technology; afterward, it was an exporter.[9] Thomas Cochran's *Frontiers of Change* contains essentially the same message.[10]

This interpretation of American technology has derived in large part from historians' excessive attention to the so-called American system of manufactures. A wealth of literature on this topic, including my own contribution, has served to obscure even larger patterns of technological development in the United States. Let me be more specific; historians have developed their overarching interpretation of American technology largely out the special reports of Joseph Whitworth, George Wallis, and the Committee on the Machinery of the United States of America led by John Anderson.[11] The history of historical writing on the American system will illuminate this point.

Duncan L. Burn, an Englishman, first focused historians' attention on these British reports in 1931 when he published his important essay, "The Genesis of American Engineering Competition, 1850–1870."[12] Burn sought not to characterize the entire history of American technology but to explain the decline of British industry and its loss of competitiveness in world markets. Readers of Burn's essay should bear in mind the context in which he wrote; Britain was in an even deeper depression than was the United States and had indeed been overrun for at least a generation by American-made products, especially consumer durables, much in the way some Japanese consumer products now dominate U.S. markets.[13] Burn's essay centered on the relative sizes of the British and American markets and how market size shaped production technologies in the two countries. In writing his essay, Burn unearthed the Whitworth and Wallis reports. More than two decades later, these reports would become basic to John E. Sawyer's now-familiar article, "The Social Basis of the American System of Manufacturing."[14]

Published in 1954, Sawyer's essay has probably shaped our understanding of the history of technology in the United States as much or more than any other single piece of scholarship.[15] Yet a critical review reveals it to be at once highly nationalistic and extremely short on historical analysis. Sawyer did little historical research beyond the Whitworth and Wallis reports; he constructed his essay principally out of these Englishmen's views on American manufacturing and the then-contemporary reports of British delegations studying productivity in the United States during the period after World War II. For readers in 1954, what Sawyer lacked in scholarly rigor was made up for in self-indulgent nationalism. His essay fit perfectly into the prevailing ideology of the United States's political, moral, and technological superiority and suggested that there was definitely an "American technology."

Less than a decade later, H. J. Habakkuk published his landmark study, *American and British Technology in the Nineteenth Century*.[16] Although clothed in the mantle of sophisticated neoclassical economic analysis, Habakkuk's book paralleled Sawyer's earlier article; it too accepted the Whitworth and Wallis reports at face value and failed to go much beyond them with additional historical research. In many respects, Habakkuk's study depended on the claim that Americans readily adopted labor-saving inventions whenever possible owing both to social factors and especially the relative dearness of labor in the United States.

After the appearance of this book, virtually all scholarship on technology in nineteenth-century America was preoccupied with Habakkuk's analysis, particularly on the problems of labor saving and of factor substitution models.[17] Implicit in all such scholarship was the idea that because it mechanized its manufacturing processes, the United States was technologically superior to Great Britain. Subsequent publication by Nathan Rosenberg of the Whitworth, Wallis, and Anderson reports and his *Technology and American Economic Growth*, while evidencing impressive historical research and analysis, merely reinforced the already dominant view that the manufacture of consumer durables somehow characterized all of technology in nineteenth-century America.[18]

Sawyer's article resonated with other historical trends of the early 1950s particularly the growth of the American studies movement and the prolific but short-lived Center in Entrepreneurial History at Harvard. Although the American studies movement can be traced to the publication of Vernon L. Parrington's *Main Currents in American Thought* and Perry Miller's corpus of works on the New England mind, the appearance of Henry Nash Smith's *Virgin Land* in 1950 opened up the field and established the myth-and symbol approach that until recently has dominated American studies scholarship.[19] Soon American studies scholars such as Marvin Fisher, John Ward, Leo Marx, and Alan Trachtenberg were publishing works dealing in one way or another with technology in the United States.[20] Because they focused strictly on the United States, these works implicitly suggested that the American experience was both unique and superior. John Kouwenhoven's *Made in America* moved beyond myth and symbol by treating American vernacular building, but like the myth-and-symbol works, it failed to look beyond the shores of the United States before characterizing what was uniquely "American."[21] More recently, John Kasson grafted the ideas of Bailyn and others about republican ideology onto the mainline American studies interpretations of technology in the United States to arrive at *Civilizing the Machine,* a more mature book but one not entirely free of old-style nationalism.[22]

Although its focus was different, Arthur Cole's Center in Entrepreneurial History shared the same nationalist tenor as American studies. Hugh G. J. Aitken, Cole's protégé and a major contributor to the field of the history of technology, recently stated that Cole's enterprise was highly chauvinistic.[23] Scholars shared a view that the United States was a perfect place and that entrepreneurial history would help the rest of the world replicate the American economy, its business, and its entrepreneurship. Cole's center folded in 1958 with the loss of funding and Cole's retirement, but the scholars who had been involved continued to flourish, just as did the American studies movement.

At the same time, American historians such as Daniel Boorstin and David Potter were writing "feel-good" history. Boorstin's *The Americans: The National Experience* and especially his *The Americans: The Democratic Experience* recounted many stories of Yankee ingenuity.[24] In the tradition of Frederick Jackson Turner, Potter's *People of Plenty* focused on the abundance of resources and the American economy as factors in making the American experience entirely unique.[25] Neither Boorstin nor Potter based his interpretation of the American experience on any comparative data. The same was true of most American history written in the 1950s and 1960s. When a nation is the spiritual, moral, economic, political, and technological leader of the world, its historians need not search beyond its boundaries for its inspiration.

But beginning in the late 1960s and continuing through the 1970s, the consensus in American historiography began to fall apart. "Darkside" historians, to use Brooke Hindle's expression,[26] began to write a new social history of the American experience. Inspired by British and Continental historians such as E. P. Thompson and Fernand Braudel, these historians began to see a different picture. World events—many of them technological in nature, from the launching of Sputnik to the use of napalm in Vietnam to the Japanese invasion of American markets—helped to fuel this new history. Paradoxically, historians of American technology have been among the last to see the significance of the new social history just as they have largely missed the significance of the work by Bailyn and others. To be sure, they have begun to look at social conflict as well as other aspects of the "dark side" such as feminism and the labor process,[27] but they have not abandoned the overarching nationalist interpretation of technology in America.

If historians of American technology are to move significantly beyond consensus history, they must seek a more internationalist context.[28] They must also free themselves from placing too great an emphasis on factor substitution models, which have served to focus historians' attention too narrowly on "labor-saving" technology. This narrow focus has led them to ignore major areas of technological development in the United States. Some historians, however, have already begun to overcome these problems.

Newer scholarship on the American system of manufactures itself has raised questions about traditional assumptions. One of these questions concerns the "American-ness" of the American system. Certainly there were important European precedents, such as the work of Christopher Polhem in Sweden, Honoré Blanc in France, and several engineers in England.[29] The ambiguity of the so-called American system also has been emphasized recently.[30] Likewise, assumptions about the social factors shaping the development of American manufacturing technology have been called into question, such as the idea that the willingness of Americans to buy standardized goods allowed manufacturing processes to be mechanized whereas more discriminating European consumers demanded custom-made goods requiring extensive hand processes.[31] Finally, the presumption of a rapid diffusion of the American system beyond the confines of the military-supported arms factories has been questioned.[32]

Ultimately, however, the large amount of scholarship on the American system may in fact be out of proportion to its overall importance to the economic and technological development of the United States relative to other, badly neglected areas. One can think of a number of such topics: the development of iron and steel technology; technologies that made possible the rise of the modern city (civil engineering, building, transportation, electrical communications, power and light, etc.); petroleum exploration, production, and refining technologies; the development of the chemical industry and its technologies; the development and diffusion of the internal combustion engine and the steam turbine; printing technologies; textile technologies after 1850, including the rise of man-made fibers; the machine-tool industry and its products; refrigeration; the production and processing of foodstuffs; industrial safety; and technical education (including the development of engineering theory). Moreover, newer themes need to be better developed such as the role of government, the military, and noneconomic factors in the development of technology in the United States. This is but a partial list, intended to provoke rather than be comprehensive.

Comparative history offers one means to move away from a highly chauvinistic brand of scholarship. But comparative studies are few in number. David Jeremy's *Transatlantic Industrial Revolution* is not explicitly a comparative history, but certainly its strength in interpreting the development of antebellum U.S. textile technology derives from its thorough grounding in English developments.[33] Thomas P. Hughes's *Networks of Power* is unquestionably the most significant comparative study published to date, but even this important work is devoted as much to arguing for the primacy of technological systems as it is to understanding the comparative histories of electrification in the United States, Germany, and England.[34] Cecil Smith's as-yet-unpublished paper on steam boiler explosions in Germany, France, England, and the United States clearly demonstrates the value of comparative history.[35] Other historians of technology should follow the lead of Jeremy, Hughes, and Smith.

Scholarship in the history of technology published in the last five years by a number of non-American scholars has raised important questions about the sources of American technology in the period after 1851. Hans-Joachim Braun's work is highly provocative in this respect. Braun has traced a fascinating network of German-American technologists who transfered technology and technical information between Germany and the United States from 1884 to 1930. Although transfers went in both directions, one is particularly impressed by the amount and nature of the technology flowing from Germany to the United States. In refrigeration, steam pumps, steam engines and turbines, injectors, coke ovens, gasholders, and many areas of mechanical engineering, German expertise proved critical to technological development in the United States.[36]

Braun has begun to broaden his work beyond the formal National Association of German-American Technologists, and with effect. He has demonstrated a significant flow of German and Swiss power technologies to the United States between 1880 and 1939.[37] Germany was, after all, the home of the Otto and Langen gas engine, the Otto four-stroke engine, and the Diesel engine.[38] I believe that if historians were to look carefully, they would see important European developments in such areas as precision measuring instruments and machine tools and heavy-equipment design and manufacture.

The list of other German contributions to American technology is lengthy: Germany trained many of the mining engineers of the American West, and certainly the Columbia School of Mines was modeled after the great Bergakademie Freiberg (and the French École des Mines).[39] Mechanical refrigeration was a German development, which stemmed from the more rigorous theoretical bent of German technical education.[40] Press working of sheet steel into a wide range of components for consumer durable goods also appears to have been imported from Germany in the last quarter of the nineteenth century.[41] This as-yet-anonymous technology has been fundamental to the automobile and many other industries of the United States. Germany's organic chemicals industry totally dominated the world's markets by the time of World War I. With the war's outbreak, both England and the United States faced uphill struggles to create domestic industries that could provide much needed dyestuffs and pharmaceuticals.[42] After investing tens of millions of dollars in a fledgling dyestuffs industry, Du Pont's executives determined in 1920 that the best solution for the company was to import German dye chemists. Severely criticized for this move, company president Irénée du Pont argued,

> After two years [of] work in the development of the dye industry we felt sure that many needless experiments could be avoided if there were available in this country men who had practical experience in the dye industry. . . . I have no doubt that American chemists in time can solve the same problems which were solved by the German chemists. Neither Germany nor the United States has a monopoly in brains, but there is a grave economic waste, both in money and time, in slowly and laboriously performing over again experiments which have already been made.[43]

Almost two decades earlier, Du Pont, like General Electric and Kodak, had established its first research and development laboratories, modeling them after those run by the German chemical industry. Furthermore, the Germans taught the American chemical industry much about industrial safety, safety explosives in mining, and industrial toxicology.[44] Du Pont was certainly not the only American chemical company to employ German chemists.

Nor was Germany the United States's only technological mentor in the period after 1851. The transfer of the British Bessemer steel process to the United States and its subsequent modification in the American environment are well known.[45] Recently, Geoffrey Tweedale has written convincingly about the continued superiority of Sheffield steel in the United States's markets in the second half of the nineteenth century and early twentieth century. Despite efforts to match Sheffield's famed steel—including the importation of workers and equipment from Sheffield—the United States never matched the output or quality of the English tool steel. Perhaps more important, England remained a metallurgical leader through research. Such efforts led to the development of alloy steels including those made with manganese, tungsten, vanadium, and chromium.[46] Although the United States steel industry might have been the largest in the world by 1900, it lacked the capabilities of the British in the "specialty" end of the business.

France also made its contributions to the development of technology in the United States. French contributions to technical education in the United States have been well documented.[47] Edwin T. Layton, Jr., has recently cited the importance of French engineering theory to the development of various technologies in the United States.[48] In his forthcoming book, Layton documents how the turbine came to the United States from France, was modified by Americans using both French and homegrown theory and American-derived experimental data, and then was modified again in Germany based on a theory developed by German engineers but using American test data.[49] James Flink also discusses French contributions in his forthcoming world history of the automobile. He notes that the automobile was essentially a French invention, although American manufacturers determined how to produce them in abundance.[50] The French were also among the pioneers in the development of the first man-made fibers; cellophane, which long earned sizable profits for Du Pont, was also a French invention.[51]

Other nations likewise contributed greatly to American technology. The manufacture of the Ford Model A would not have been possible without Henry Ford's purchase of the Johansson Gauge Company of Sweden and precision measuring machines from Switzerland.[52] American hydraulic engineering greatly depended on European know-how in turbine design and execution; the first turbines for the Niagara Falls hydroelectric complex were, after all, of Swiss design.[53] Electrochemistry flourished in certain areas of Europe, and much of the knowledge of this new technology was transferred to the United States.[54] Sanitary engineering in the United States was based in large measure on European ideas.[55] European developments figure importantly in practically any area to which one turns, yet one seldom finds such developments discussed in general works on American technology. American historians have complacently assumed both a superiority and insularity of "their" technology that has little basis in fact. The same might be said, it should be noted, of the histories of other nations' technology.[56]

The United States has been called "a nation of nations."[57] Immigration of Europeans to the United States in the second half of the nineteenth and the early part of the twentieth century was a major development in American history. Yet outside of the attention paid to the best known "immigrant inventors" such as Michael Pupin and Charles P. Steinmetz, historians of technology in America have largely ignored immigration as a source of American engineering know-how and skilled labor.[58] Although perhaps a small fraction of the total immigrant population, skilled engineers and workers certainly played an important role in the development of technology in the United States. Thomas Carroll has provided evidence that a large number of European chemists immigrated to the United States from the last third of the nineteenth century to the outbreak of World War II, and many of them worked in American industry.[59] The development of manufacturing technology at the Ford Motor Company in its first two prolific decades would certainly have been different without European-trained machinists such as Oscar Bornholdt and Carl Emde, among others.[60] In the Du Pont Company, a German-trained chemist from Norway, Fin Sparre, played a critical role in charting the company's diversification strategy, in following international developments in the chemical industry, and in negotiating and monitoring a major patents and processes agreement with Imperial Chemical Industries under which an enormous amount of scientific and technical information was formally and informally exchanged between the two chemical giants.[61] Similar contributions must have been made to other American companies. Historians should borrow some of the techniques of the "new history" to assess the number and activities of these immigrant technologists from Europe. Collective biography or prosopography might be one approach, but even the assemblage of basic data on immigrant technologists would be helpful.

A large number of pure-bred American engineers and mechanics took tours of Europe to learn more about new technolgies there. Thanks to Darwin Stapleton, we have at least a vague idea about the period before the Civil War, but our ignorance grows exponentially in the years after.[62] Many American scientists, some of whom entered the employ of industry, obtained graduate and postgraduate training in Europe. We need better information about their education and subsequent contributions.

To understand fully the development of technology in the United States, we must also assess the role of engineering theory. Here, too, Europe looms large. In thermodynamics, mechanics, hydrodynamics, aerodynamics, electrodynamics, signal theory, statics, and other areas of the "engineering sciences," European engineers provided much of the basic work in the era commonly believed to mark the beginning of the "American century." Historians of technology in the United States need to develop a much more comprehensive picture of the development of the engineering sciences and then fit this history with the overall pattern of technological development in the United States.

Even if we were to assemble all of the historical data for which I have called, we are still left with the question of whether there is indeed an "American technology." Technology is a human phenomenon that is not confined within national boundaries. Unquestionably there are, however, important differences in the technologies of different countries; toilets in European *are* different from those in the United States. The economist might use a factor substitution model to explain those differences; the anthropologist is likely to point to the cultural differences as determinants; the cultural geographer might look to geographical and demographic factors. With some important exceptions, historians writing about the development of technology in the United States particularly after 1860 have been so prone to assume American priority and superiority that they have largely ignored developments in Europe. Often, their claims of distinction have had no comparative basis. Studying European developments would allow us to raise the analysis of American technology to a higher level and to achieve a much richer textured history. We must above all abandon the old-style nationalism that has distorted so much of our historical scholarship. By developing a history of American technology based on comparative analysis we will ultimately achieve an improved understanding of cultural and economic variation in technology.

Melvin Kranzberg's career—especially his founding of *Technology and*

Culture, the *international* journal of the Society for the History of Technology—has been devoted to building a more comprehensive understanding of the development of technology and its role in society. Kranzberg's training as a European historian has surely served the Society for the History of Technology well in helping to keep a narrow parochialism at bay. Under different leadership, *Technology and Culture* might easily have slipped into being a journal celebrating the achievements of American technology. The next generation of historians of technology should strive even harder to build a more complete, less nationalistic understanding of the history of technology.

NOTES

I wish to thank scholars at the University of Toronto and Harvard University for their criticism of orally presented versions of this paper and especially Charles Haines and Steven Usselman for their critical reading of an initial draft.

1. Brooke Hindle, "British v. French Influence on Technology in the Early United States," in *Actes du XIᵉ Congrès International d'Histoire des Sciences, 1965* 6 (Warsaw: Polish Academy of Sciences, 1968), pp. 49–53.

2. Frederick Jackson Turner, "The Significance of the Frontier in American History," in *Annual Report of the American Historical Association* (Washington, D.C.: Government Printing Office, 1893), pp. 199–227.

3. Bernard Bailyn, *The Ideological Origins of the American Revolution* (Cambridge, Mass.: Harvard University Press, 1967); Gordon Wood, *The Creation of the American Republic, 1776–1787* (Chapel Hill: University of North Carolina Press, 1969); Joyce O. Appleby, *Capitalism and a New Social Order* (New York: New York University Press, 1984).

4. Appleby, *Capitalism and a New Social Order,* p. 7.

5. Rhys Isaac, *The Transformation of Virginia, 1740–1790* (Chapel Hill: University of North Carolina Press, 1982).

6. "Engines of Change: An Exhibition on the American Industrial Revolution," and Brooke Hindle and Steven Lubar, *Engines of Change: The American Industrial Revolution 1790–1860* (Washington, D.C.: Smithsonian Institution Press, 1986).

7. *Engines of Change,* p. 268.

8. Charles T. Rodgers, *American Superiority at the World's Fair* (Philadelphia: J. J. Hawkins, 1852).

9. Nathan Rosenberg, *Technology and American Economic Growth* (New York: Harper and Row, 1972).

10. Thomas C. Cochran, *Frontiers of Change* (New York: Oxford University Press, 1981). Other scholarship also carries a similar theme. See, e.g., Brooke Hindle, *Technology in Early America* (Chapel Hill: University of North Carolina Press, 1966) and George H. Daniels, "The Big Questions in the History of American Technology," in *The State of American History,* ed. Herbert J. Bass (Chicago: Quadrangle, 1970), pp. 197–219. Eugene S. Ferguson, "The American-ness of American Technology," *Technology and Culture* 20 (January 1979): 3–24, also discusses the Crystal Palace exhibition but argues that American "missionary zeal" had existed long before 1851.

11. Nathan Rosenberg, ed., *The American System of Manufactures: The Report of the Committee on the Machinery of the United States 1855 and the Special Reports*

of George Wallis and Joseph Whitworth 1854 (Edinburgh: Edinburgh University Press, 1969).

12. D. L. Burn, "The Genesis of American Engineering Competition, 1850–1870," *Economic History* 2 (1931): 292–311.

13. One can draw a crude parallel between Burn's motivation in explaining the decline of British industry in the nineteenth century in the face of competition from the United States and the wealth of current literature seeking to explain the decline of U.S. competitiveness and the rise of Japanese manufacturing.

14. John E. Sawyer, "The Social Basis of the American System of Manufacturing," *Journal of Economic History* 14 (Winter 1954): 361–79.

15. Many of the standard bibliographies in the history of technology treat Sawyer's essay as being important. See, e.g., Eugene S. Ferguson, *Bibliography of the History of Technology* (Cambridge, Mass.: MIT Press, 1968), p. 299; Brooke Hindle, *Technology in Early America*, p. 48; and Marc Rothenberg, *The History of Science and Technology in the United States: A Critical and Selective Bibliography* (New York: Garland, 1982), p. 200.

16. H. J. Habakkuk, *American and British Technology in the Nineteenth Century* (Cambridge: Cambridge University Press, 1962).

17. For a review of this literature, see Paul Uselding, "Studies of Technology in Economic History," in *Recent Developments in the Study of Economic and Business History: Essays in Memory of Herman E. Krooss,* ed. Robert E. Gallman (Greenwich, Conn.: JAI Press, 1977), pp. 159–219.

18. See nn. 9 and 11.

19. Vernon L. Parrington, *Main Currents in American Thought,* 3 vols. (New York: Harcourt, Brace, & Co., 1927–30); Perry Miller, *The New England Mind: The Seventeenth Century* (New York: Macmillan, 1939); Henry Nash Smith, *Virgin Land: The American West as Symbol and Myth* (Cambridge, Mass.: Harvard University Press, 1950). The literature on the origins and development of American studies is abundant. I have found Gene Wise, " 'Paradigm Dramas' in American Studies: A Cultural and Institutional History of the Movement," *American Quarterly* 31 (1979): 293–331, particularly helpful.

20. Marvin M. Fisher, *Workshops in the Wilderness: European Response to American Industrialization, 1830–1860* (New York: Oxford University Press, 1967); idem, "The Iconology of Industrialism, 1830–1860," *American Quarterly* 13 (Fall 1961): 347–64; John William Ward, "The Meaning of Lindbergh's Flight," *American Quarterly* 10 (Spring 1958): 3–16; Leo Marx, *The Machine in the Garden* (New York: Oxford University Press, 1964); Alan Trachtenberg, *Brooklyn Bridge: Fact and Symbol* (New York: Oxford University Press, 1965).

21. John A. Kouwenhoven, *Made in America: The Arts in Modern Civilization* (Garden City, N. Y.: Doubleday, 1948).

22. John F. Kasson, *Civilizing the Machine: Technology and Republican Values in America, 1776–1900* (New York: Grossman, 1976). In 1985, a number of European scholars severely criticized U.S. American studies scholars for their parochialism. See K. J. Winkler, "Scholars Chide American Studies for Ignoring the Rest of the World," *Chronicle of Higher Education* 31 (13 November 1985): 7–8.

23. Hugh G. J. Aitken, "Engineers and Entrepreneurs," paper delivered at the Annual Meeting of the Society for the History of Technology, Pittsburgh, Penn., 25 October 1986. For more information on Cole and his center, see Steven A. Sass, *Entrepreneurial Historians and History: Leadership and Rationality in American Economic Historiography, 1940–1960* (New York: Garland, 1986).

24. Daniel J. Boorstin, *The Americans: The National Experience* (New York: Random House, 1965), and idem, *The Americans: The Democratic Experience* (New

York: Random House, 1973). See also Boorstin's earlier *The Americans: The Colonial Experience* (New York: Random House, 1958).

25. David M. Potter, *People of Plenty: Economic Abundance and the American Character* (Chicago: University of Chicago Press, 1954).

26. Brooke Hindle, " 'The Exhilaration of Early American Technology': A New Look," in *The History of American Technology: Exhilaration or Discontent?*, ed. David A. Hounshell (Wilmington, Del.: Hagley Museum and Library, 1984), p. 13.

27. See the discussions by Hindle, Carroll W. Pursell, Jr., and Stuart W. Leslie in *The History of American Technology*.

28. This is essentially the argument made by Thomas P. Hughes, "Emerging Themes in the History of Technology," *Technology and Culture* 20 (October 1979): 697–711.

29. David A. Hounshell, *From the American System to Mass Production, 1800–1932* (Baltimore: Johns Hopkins University Press, 1984), pp. 24–27; William A. Johnson, trans., *Christopher Polhem: The Father of Swedish Technology* (Hartford, Conn.: Trinity College, 1963), pp. 109–35; Carolyn C. Cooper, "The Production Line at Portsmouth Block Mill," *Industrial Archaeology Review* 6 (Winter 1981–82): 28–44; idem, "The Portsmouth System of Manufacture," *Technology and Culture* 25 (April 1984): 182–225.

30. Hounshell, *From the American System*, pp. 17–25.

31. The best example is the performance of the Singer Manufacturing Company in Europe in the nineteenth century. See ibid., pp. 82–121, and Robert Bruce Davies, *Peacefully Working to Conquer the World: Singer Sewing Machines in Foreign Markets, 1854–1920* (New York: Arno, 1976). Singer found it easier to sell sewing machine cabinets made from the inferior gum wood in Europe than it did in the United States, where consumers expected high grades of walnut, cherry, and oak. See Hounshell, *From the American System*, pp. 140, 142.

32. Hounshell, *From the American System*, pp. 67–215, passim. But compare Donald Hoke's interpretation in "Ingenious Yankees: The Rise of the American System of Manufactures in the Private Sector" (Ph.D. diss., University of Wisconsin, 1984).

33. David J. Jeremy, *Transatlantic Industrial Revolution: The Diffusion of Textile Technologies between Britain and America, 1790–1830s* (Cambridge, Mass.: MIT Press, 1981).

34. Thomas P. Hughes, *Networks of Power: Electrification in Western Society, 1880–1930* (Baltimore: Johns Hopkins University Press, 1983).

35. Cecil O. Smith, "Making Steam Safe: The Regulatory Response to Boiler Explosions in France, Germany, Britain, and the USA, 1830–1920," paper presented to the HOT Lunch Seminar, University of Delaware, 18 April 1985.

36. Hans-Joachim Braun, "The National Association of German-American Technologists and Technology Transfer between Germany and the United States, 1884–1930," *History of Technology* 8 (1983): 15–35.

37. Hans-Joachim Braun, "The Adaptation of German and Swiss Power Technologies in the United States, 1880–1939," paper given at the annual meeting of the Economic History Association, Hartford, Conn., September 1986.

38. On German contributions to the development of the internal combustion engine, see the corpus of works by Lynwood Bryant: "The Beginnings of the Internal Combustion Engine," in *Technology in Western Civilization*, ed. Melvin Kranzberg and Carroll W. Pursell, Jr. (New York: Oxford University Press, 1967), pp. 648–64; "The Silent Otto," *Technology and Culture* 7 (Spring 1966): 184–200; "The Origin of the Four-Stroke Cycle," *Technology and Culture* 8 (April 1967): 178–98; "Rudolph Diesel and his Rational Engine," *Scientific American* 221 (August 1969): 108–18;

"The Role of Thermodynamics in the Evolution of Heat Engines," *Technology and Culture* 14 (April 1973): 152–65; "The Development of the Diesel Engine," *Technology and Culture* 17 (July 1976): 432–46; "The Internal Combustion Engine," in *A History of Technology*, ed. Trevor I. Williams (Oxford: Oxford University Press, 1978), vol. 7, pt. 2, pp. 997–1024. See also C. Lyle Cummins, Jr., *Internal Fire* (Lake Oswego, Oreg.: Carnot Press, 1976). The rich German context of Rudolph Diesel's work is treated in Donald E. Thomas, Jr., *Diesel: Technology and Society in Industrial Germany* (University: University of Alabama Press, 1987), pp. 38–67.

39. Before 1865, at least seventy-five Americans studied at Freiberg. Eleven studied at the École des Mines. See Thomas T. Read, *The Development of Mineral Industry Education in the United States* (New York: American Institute of Mining and Metallurgical Engineers, 1941), pp. 26–29. On the founding of the Columbia School of Mines, see James Kip Finch, *A History of the School of Engineering, Columbia University* (New York: Columbia University Press, 1954), pp. 3, 27–41. See also Clark C. Spence, *Mining Engineers and the American West: The Lace-Boot Brigade, 1849–1933* (New Haven: Yale University Press, 1970), pp. 1–39, passim, for a discussion of the European training of American mining engineers.

40. Oscar E. Anderson, *Refrigeration in America: A History of a New Technology and Its Impact* (Princeton: Princeton University Press, 1953), pp. 71–85, treats the European background of mechanical refrigeration. On the refrigeration work of the German engineer/educator Carl P. G. R. Linde, see Linde, *Aus meinem Leben und von meiner Arbeit* (Munich: R. Oldenbourg, 1916).

41. Hounshell, *From the American System*, pp. 209 and 369 n. 71. Although not dealing explicitly with the press working of metals, two recent papers indicate the significant number of skilled machinists and metal workers in the areas where press working was developed. See Hartmut Keil, "Chicago's German Working Class in 1900," and John B. Jentz, "Skilled Workers and Industrialization: Chicago's German Cabinetmakers and Machinists, 1880–1900," in *German Workers in Industrial Chicago, 1850–1910: A Comparative Perspective,* ed. Hartmut Keil and John B. Jentz (DeKalb, Ill.: Northern Illinois University Press, 1983), pp. 19–36, 73–85.

42. See Williams Haynes, *The American Chemical Industry* (New York: Van Nostrand, 1945), vol. 2.

43. Quoted in David A. Hounshell and John K. Smith, *Science and Corporate Strategy: Du Pont R & D, 1902–1980* (New York: Cambridge University Press, 1988), chap. 3.

44. Ibid., chap. 1 and 24. Both Du Pont and the Bureau of Mines purchased safety explosive testing equipment from the Germans, and the Germans were about two decades ahead of the American chemical industry in their knowledge of industrial toxicology.

45. Jeanne McHugh, *Alexander Holley and the Makers of Steel* (Baltimore: Johns Hopkins University Press, 1980), is the standard work.

46. Geoffrey Tweedale, "Metallurgy and Technological Change: A Case Study of Sheffield Specialty Steel and America, 1830–1930," *Technology and Culture* 27 (April 1986): 189–222. See also idem, *Sheffield Steel and America: A Century of Commercial and Technological Interdependence 1830–1930* (Cambridge: Cambridge University Press, 1987).

47. See, e.g., Peter M. Molloy, "Technical Education and the Young Republic: West Point as America's École Polytechnique, 1802–1833" (Ph.D. diss., Brown University, 1975); Bruce Sinclair, "The Promise of the Future: Technical Education," in *Nineteenth Century American Science,* ed. George H. Daniels (Evanston, Ill.: Northwestern University Press, 1972), p. 260. Karl-Heinz Manegold, "Technology Academized: Education and Training of the Engineer in the Nineteenth Century," in *The Dynamics of Science and Technology: Social Values, Technical Norms and*

Scientific Criteria in the Development of Knowledge, ed. Wolfgang Krohn et al., (Dordrecht: Reidel, 1978), pp. 141–42, discusses the influence of the École Polytechnique on German technical education.

48. Edwin T. Layton, Jr., "European Origins of the American Engineering Style of the Nineteenth Century," in *Scientific Colonialism: A Cross-Cultural Comparison,* ed. Nathan Reingold (Washington, D.C.: Smithsonian Institution Press, 1987), pp. 151–66.

49. Personal communication, Edwin T. Layton, Jr., to David Hounshell, 16 April 1986.

50. James J. Flink, *The Automobile Age* (Cambridge, Mass.: MIT Press, 1988). See also idem, "Entrepreneurship in the Automobile Industry," paper given at the annual meeting of the Society for the History of Technology, Pittsburgh, Penn., 25 October 1986.

51. On the transfer of French man-made fibers and cellophane technology to the United States, see Hounshell and Smith, *Science and Corporate Strategy,* chap. 8. Du Pont also transferred the French-developed Claude process for the manufacture of ammonia; see ibid., chap. 9. Cf. the transfer of British rayon technology to the United States as detailed in D.C. Coleman, *Courtaulds: An Economic and Social History* (Oxford: Oxford University Press, 1969), vol. 2, pp. 104–19.

52. On Johansson and Ford, see Hounshell, *From the American System,* p. 286, and Torsten K. W. Althin, *C. E. Johansson, 1864–1943* (Stockholm: Privately printed, 1948). My research in the Ford Archives, Edison Institute, Dearborn, Mich., revealed significant purchases of Swiss precision measuring machines at the time of the changeover to the Model A. See particularly Accessions 38 and 390.

53. Edward Dean Adams, *Niagara Power: History of the Niagara Falls Power Company, 1886–1918* (Niagara Falls, N.Y.: Privately printed, 1927) 2: 439. Robert Belfield, "The Niagara System: The Evolution of an Electric Power Complex at Niagara Falls," *Proceedings of the IEEE* 64 (September 1976): 1344–50, interprets the Niagara Falls complex as a synthesis of best-practice European and American technology.

54. Although still nationalistic, Martha Moore Trescott, *The Rise of the American Electrochemical Industry, 1880–1910: Studies in the American Technological Environment* (Westport, Conn.: Greenwood, 1981), pp. 15–17, offers a more balanced view than that of Haynes's *The American Chemical Industry.*

55. Joel Tarr, et al., "Water and Wastes: A Retrospective Assessment of Wastewater Technology in the United States, 1800–1932," *Technology and Culture* 25 (April 1984): 234.

56. See, e.g., Ulrich Troitsch and Wolfhard Weber, eds., *Die Technik: von den Anfängen bis zur Gegenwart* (Braunschweig: Westermann, 1982).

57. Walt Whitman wrote, "These States are the amplest poem, / Here is not merely a nation, / but a teeming nation of nations." Quoted from Peter C. Marzio, ed., *A Nation of Nations* (New York: Harper and Row, 1976), forematter epigraph.

58. Most of the literature on immigration and ethnicity also ignores engineers and highly skilled labor. *German Workers in Industrial Chicago* is an exception.

59. P. Thomas Carroll, "Academic Chemistry in America, 1876–1976: Diversification, Growth, and Change" (Ph.D. diss., University of Pennsylvania, 1982), pp. 248–302. We need similar studies for other academic disciplines related to engineering and technology.

60. Hounshell, *From the American System,* pp. 223–30, 241, 247–48, 270–71.

61. Hounshell and Smith, *Science and Corporate Strategy,* chap. 2, 3, 10, 16.

62. Darwin H. Stapleton, *Accounts of European Science, Technology, and Medicine Written by American Travelers Abroad, 1735–1800, in the Collection of the American Philosophical Society* (Philadelphia: American Philosophical Society, 1985).

Historians of Technology and the Context of History

BROOKE HINDLE

The recent rise of the history of technology as a profession has far deeper cause and far more expansive meaning than is suggested by merely identifying our many friends, most of them still living, who are responsible for this great development. The history of technology is a fundamental and growing dimension of history as a whole. It had to grow and must continue to grow because technology is so important in today's world. Its history must therefore be increasingly integrated into our understanding of human history as a whole.

Most history books and most history taught in schools still include only minimal attention to technology. Because historians of technology are convinced of its significance, many complain almost automatically of the inadequate attention given to it in general history. There are few generalists, however, and historians specializing in other areas cannot be expected to give the history of technology adequate attention in their own histories. Some historians have increased their attention to technology, but the needed integration has to rest most heavily on historians of technology, a number of whom recognize the need. They have moved to close the gap but sometimes remain too separated from history as a whole to be fully effective.

General history has been changing and developing too; some of its problems and characteristics are shared by the history of technology while others are not. Indeed, the multiplying areas of specialization within history have their own differences, although most share with the history of technology an inadequate and often decreasing sensitivity to general history and to the enormous social need for working with and within large-scale historical syntheses.

The greatest role of history is as our social memory. Over the years, historians have concerned themselves with such critical questions as whether history is a science or an art, a humanity or a social science, and, more sensitively, whether history should focus more upon facts or upon interpretation. These are serious matters that remain not fully resolved. They

230

are, however, far less important than the constant and unavoidable function of history as society's memory.

This role is best understood as a direct parallel to the personal memory of every individual. Humans cannot function except on the basis of personal memory—over which they exercise only minor control. Individuals may forget (or push into the back of their minds) unpleasant experiences but, at the same time, exaggerate cherished events. Beyond the individual's control, however, memory fades and grows in differing directions—usually strengthening the positive and weakening the negative. Yet most of what is remembered reflects experiences seen and felt, perpetuating the biases with which they were originally viewed.

If an individual intentionally or unintentionally reverses this pattern of memory, by emphasizing and exaggerating negative experiences and letting positive experiences fade, the likely result is a depression and an inability to function effectively. Reasonably satisfactory human life requires that the memory be primarily positive, that it emphasize past capabilities and clearly recall deficiencies that need to be improved. Remembering the past as nothing but defeats and failures is a route to disaster.

Society has to use its memory—its history, that is—in the same manner. Society's history points toward its future. In fact, "history is a dialogue in the present with the past about the future."[1] Of course, accuracy and precision are required in the history used socially, or it will lead to defeats. Yet, even accurate history, like reliable personal memory, can vary enormously through the factual data collected and organized and, even more, through the patterns of interpretation applied.

The history used by many elements of society differs widely from that created, taught, and published by professional historians. It is much more diverse and individualistic in character, because "every man" is indeed "his own historian."[2] Of course, some individuals and groups are much more influenced than others by professional history. Whatever the degree of influence, however, everyone's sense of the present and future state of society rests heavily on a personal historical understanding, whether superior or minimal.

Historians are primarily governed by the tenets of their profession although most of us have some sensitivity to popular or general social concepts of history and to the need to communicate the best professional history to society. Even so, there is still too little recognition that if professional historians do not supply well-researched history in the popular areas and of the character needed by social groups, people will find what they need or make it up on their own. There are times in specific historical areas when society tends to walk a different path from that being followed by professional historians.

This divergence between professional history and the interest and beliefs of society at large is increasing at this time. Among the causes, the most obvious may be differences in attitudes and aspirations, but one that deserves particular attention is the rapid increase of specialization and the decrease of large-scale historical syntheses among professionals. Syntheses are particularly important to the general public because they offer a sense of where we are and how we have gotten here; they even seem to give an idea of what is possible for the future. The rise of specialization is directly related to the decrease of professional syntheses.

The decline in the publication of large syntheses, such as Charles Beard's American histories, is obvious, but the situation in the teaching of history is even more dramatic. College history, first of all, declined precipitately after the student rebellions of the mid- to late 1960s until it began to stabilize just a few years ago. One cause of the drop in undergraduate enrollments was the widespread elimination of almost all history-course requirements. Another, however, was the elimination of traditional syntheses such as Western Civilization. A general agreement arose that Western Civilization was a bad approach because the third world, and even the second, ought not be denigrated by omission from the broadest college syntheses. An obvious answer, promoted by such historians as William McNeill, was to offer world history instead.[3] So far that effort has received very little response. American history, on the other hand, faired fairly well after an initial decline.

As for the history of technology, it has flourished in professionalization as well as in growth while the bulk of history has been declining. Certain other areas of specialization have similarly developed higher standards of accuracy and rigor, but most have had a decline in student enrollment at both graduate and undergraduate levels as well as in the number of faculty members.

Many historians today complain about the problem of specialization. A recent survey conducted by the Organization of American Historians reveals the feeling "that scholarship has become increasingly specialized in subject, method, and ideology while most historians are expected to be 'generalists.'" There is broad dissatisfaction with this "'overspecialization'" that appears "mired in details without illuminating larger issues."[4]

The decline of synthesis has many causes. In part, it is a function of professionalization, because the broader any generalization is, the more incorrect and more unacceptable are the implications of many of its details. Particularly significant is the recent decline in acceptance of long-used generalizations—both by professional historians and by an influential minority of the public. The challenged and often cancelled generalizations include the concept of progress, the belief that the American population has enjoyed unmatched opportunities, and the assumption that technology has continued to improve the welfare of humanity in general. The inability to produce new generalizations that evoke broad enthusiasm has led to a rising "focus on narrowly defined topics."[5] This is a major cause of the increased proportion of specialized studies by historians and of their narrowing nature.

Two forms of specialization dominate historians. One is topical specialization, a concentration on such studies as the history of technology, science, economics, business, community life, art, and many others. This approach discourages large-scale synthesis because each topical area has its own organizational, chronological, and relational characteristics, which never match those of a general historical synthesis. For example, the history of technology has come to concentrate so heavily upon recent history, primarily twentieth- and late nineteenth-century history, that those seeking to put together a synthesis of even the immediately preceding period do not find adequate help.[6] Moreover, the history of technology pursued by Americans is primarily focused on United States history.

Indeed, the second current mode of specialization limits its concentration to a single country or a brief chronology—or both. Such concentrations may also fail to offer the help needed by larger-scale synthesists. For example, it has been asserted that publications on early American history have become so concentrated and detailed that only experts in that field can understand them.[7]

In any academic field, such problems are an almost inevitable result of good professional development. This is notably true of the sciences because scientists make no pretense of seeking to embrace the whole of science, or even of a single science such as physics, in their own understanding. They feel that syntheses of science ought to be left to less creative people—often to science writers rather than to scientists. Yet, historical synthesis is so fundamental to society that historians cannot neglect it. Historians of technology cannot leave to others the integration of technology into the larger understanding of history because technology has not begun to be understood sufficiently even by other historians let alone by journalists.

This situation is not unique to the history of technology. Other specialties, particularly those that have arisen recently, also fail to be understood adequately by historians unconnected to the specialty in question. Philip Curtin, a leading student of African history, has commented, "We have specialists in black history, women's history, and historical demography, but people outside these specialties pay little attention to their work."[8]

The rise of professionalization has specifically altered the degree and nature of specialization. Some of the early publications in the history of technology were broadly synthetic while others were narrowly specialized. Syntheses such as Roger Burlingame's tended to be journalistic and limited in the amount of research information provided.[9] Earlier publications were specialized, some of the more useful ones written by authors primarily concerned with hardware, as was Robert Woodbury and even A. P. Usher, despite his economics approach.[10]

Since then, the specialized study of aspects of the history of technology has been developed successfully at several universities, but it now has a broader outlook. A high sensitivity is encouraged to fields related to technology, such as economics, business history, science, and, where appropriate, urban, military, and social history. Significantly, the increased emphasis on social history has brought especially heavy attention to its relationships with technology.

Most current historians of technology, therefore, like to project their work as "contextual history"—that is, to present their accounts within the larger context of those fields to which they are most significantly related.[11] This approach is correct for the most generally useful studies of specialized history. It also is directly related to the development of large syntheses, although it is not sufficient for that need.

There is a growing popular need for general syntheses into which technology must be incorporated. The need also grows for more and better syntheses within the history of technology itself. Most older efforts have now come to be regarded as unsatisfactory.

The professionalization of the history of technology has been achieved at the cost of reducing the relationship of historians of technology to general history. Some of the most creative universities have removed the history of technology from history departments and placed it in more specialized departments. While this has aided faculty research by eliminating any need to teach general history courses or to keep in touch with the problems of the larger field, it has at the same time separated faculty members and their graduate students from general academic history. Many historians of technology do not today maintain their memberships in the historical organizations with the broadest coverage. They often are closer and more congenial to such organizations as the Society for Industrial Archeology or the Association of Living Historical Farms and Agricultural Museums than to the American Historical Association or the Organization of American Historians.

The causes and advantages of this pattern are far more evident than the disadvantages. It is worth recalling that many early contributors to the history of technology retain connections with general history because they did their graduate work in history departments. Melvin Kranzberg and Carroll Pursell, who edited the first (and only) large synthesis of the history of technology written in this country, both received their degrees in history departments.[12] So did Merritt Roe Smith, whose *Harpers Ferry Armory and the New Technology* was particularly successful in carrying the history of technology into the concerns of general historians.[13]

These facts by no means suggest that the organizational changes resulting from professionalization ought to be altered. They do, however, point to the continuing need for historians of technology to remain well connected with historians in a variety of fields and well informed about generalized history and large-scale synthesis. This accomplishment has become more difficult and will require specific planning.

Both synthetic history and specialized history involve another major problem of emphasis, which can be either accuracy and objectivity in collecting research data, often called "facts," or interpretation, involving intended purpose, ideology, and even advocacy.

Each of these issues, but especially that of interpretation, was answered, in ways that would have been previously unacceptable, during the late 1960s and early 1970s—the period of the student rebellions. Most of those who had lived through World War II saw little that was positive in either these rebellions or the demands of the rebels. Today, however, many aspects of society have been significantly altered in the direction of those demands. More important, history has been altered because many of the activists of the 1960s, as well as those they influenced, have moved into the academic world, including departments of history and departments responsible for the history of technology. It is, therefore, necessary to give some thought to that period of crisis.

The dissent arose following the anti-Vietnam War and civil rights movements, but it was also reacting against historical syntheses that emphasized consensus, continuity, and positive interpretations of the past. Charles Reich referred to the crisis as "an organic one" created by multiple problems: 1) disorder, corruption, hypocricy, war; 2) poverty, distorted priorities, and law-making by private power; 3) uncontrolled technology and the destruction of environment; 4) decline of democracy and liberty; 5) artificiality of work and culture; 6) absence of community; 7) loss of self.[14]

None of those asserted problems—and few others that were ever voiced—had been caused by or were directly remediable by American universities. Yet universities became the prime target of the activists. Beginning at Berkeley with unrestrained vulgarity, the attacks moved through the country and concentrated especially on occupying university buildings and preventing their normal use. They included such anti-intellectual attacks as the firebombing of both the arts and sciences and the engineering libraries at University Heights, New York University. The negative attitude expressed toward most faculty members is best analyzed as "a strong dissatisfaction with their political inertia."[15] The activists had come to feel that professors were hiding behind objective, intellectual occupations; they demanded that faculties apply themselves and their institutions to overcoming the terrible wrongs asserted.

In retrospect, the activists of the 1960s have to be understood as forerunners of some of the "darkside" scholars of the present. They began, of course, with positive goals. They accepted and sought to fulfill, in totality, the great American ideals: freedom, justice, equality, peace, democracy, and improved living standards for all. The people they assailed believed themselves just as committed to those ideals, but were convinced that not one of them could ever be totally attained. The real question and the real difference was the degree that might reasonably be sought—and there the activists found little support from university and government leaders or from the bulk of the population.

Nevertheless, the rebels associated themselves with other activists, particularly successful ones, such as those working for civil rights. They then involved themselves in several important successes. They participated in the large-scale national reaction that caused the withdrawal of U.S. forces from Vietnam. They encouraged the rise of the women's movement. They participated in the increase of individual freedom: sexual, family, and community. Some, of course, went too far during the "me decade," in the use of drugs and in other excesses. Although numbers of them moved later to individual success of their own within the corporate world that they had heavily opposed, others did not reverse their rebellious attitudes.

Within the universities, their success was great. The bulk of students were mildly favorable or neutral toward them and only a few took opposing positions. Younger faculty members tended to be responsive even though not actively supportive. Still, the rebels' quest for total freedom resulted in the elimination of virtually all course requirements except for those within the major, while new courses on such subjects as entertainment and social "rapping" began to appear. The effect on the discipline of history was disastrous, caused by the elimination of required and optional courses plus a rising feeling that the study of the past was not worth the time. The felt need was instead to escape from manifold constrictions of the past and from the status quo. At the University of California, Los Angeles, the results were typical: during the decade of the 1970s, the number of history majors dropped by fifty-two percent, graduate history students by twenty-seven percent, and history faculty members by twelve percent. The decline then stopped.[16]

Throughout, technology was a key problem to the rebels and to others who became increasingly fearful that technology was out of control. More accurately, their antagonism towards technology was directly related to the desire for total individual freedom. Whereas other limitations on freedom might be blamed on people or organizations, many of the rebels saw technology as something with an evolutionary life of its own.

The shift of Langdon Winner from the world of the rebels to the study of technology and politics is particularly pertinent because of his darkside influence on historians. Winner concedes that he was moved to his assault on technology in 1964 at the University of California, Berkeley, by Mario Savio, the prime rebel leader. He happily reports a typical Savio outburst: "There is a time when the operation of the machine becomes so odious, makes you so sick at heart that you can't take part; you can't even passively take part, and you've got to put your bodies upon the gears and upon the wheels, upon the levers, upon all the apparatus and you've got to make it stop. And you've got to indicate to the people who run it, to the people who own it, that unless you're free, the machine will be prevented from working at all."[17]

Although most readers have accepted Winner's work as successful by providing opposition to the concept of technology being out of control, he admits that he does not want "to dismiss the notion" but to ask for "ways in which the idea could be given reasonable form." He does not deny that he "begins in criticism of existing forms but aspires to the eventual articulation of genuine, practical alternatives," that is, to "a new philosophy of technology."[18]

Activist historians and scholars cannot be condemned because they pursue their inquiries for the purpose of promoting their own desired changes in society. There was a time when historians imagined that their purpose instead should be to fulfill Leopold von Ranke's aim of studying history to find "how it really was."[19] Later, the ideal of history as a science convinced some that science flourished so well because it insisted upon using the "edge of objectivity," and that history would benefit by emphasizing the same approach.[20] More recently, however, it has become clear that complete objectivity is an unattainable dream—not only in history but in science as well. All historians are bound, inescapably, by their realm of knowledge, their experiences and conditioning, and their hopes and dreams. The best that historians can do is to recognize whence they come in their approach to history. It is possible for everyone to identify some personal biases and, at the least, to acknowledge them. Yet, even in the midst of the period of rebellion, most historians continued to believe that "Objectivity is, of course, the goal at which the scholar aims."[21] Like other ideals that can never be fully attained, it is a wonderful goal on which history rests.

This ideal has not died, but advocacy historians, who often reflect attitudes of the 1960s' rebels, seem to challenge it. They may retain a full commitment to accuracy and to the effort to collect all relevant data in their inquiries. However, they make clear their intention of using history to advance their own political and social objectives.[22] Not only advocacy historians but many others as well have generally accepted the application of moral feelings to published history. In fact, after the rebellion, historians moved toward agreeing that moral judgment might be an appropriate part of their work.[23] Also, it must be understood that the history needed by society is not an unfeeling list of facts. It has to be related both to the current state of society and to the available options that are anticipated in moving into the future.

The social need for syntheses has not been well met by historians, but those who have concentrated on specialized areas of study have a keen sense of the social importance of their work. Most of the "new histories" have a clear social focus, despite their rejection of synthesis. They include "the new economic history, the new labor history, the new social history, the new urban history, the new political history, and other greater or lesser 'news' too numerous to list." They "pronounce themselves analytical, not narrative."[24]

Another "new" with broad and significant influence was the "new left," which also related easily to the rebellion. Young new-left historians rejected the postwar historical syntheses, especially the concept of consensus that denied real conflict in the American past and declined to accept the signifi-

BROOKE HINDLE

cance of American radicals in history. New-left historians were, however, sensitive to the need for a "usable past"—that is, the need for new syntheses to replace those they had rejected.[25] Although not directly related, this development paralleled alterations in and increased strength of Marxist histories, which always had had a particular concern for the history of technology. Technology is also an interest of such new-left publications as the *Radical History Review.*

Of course, a prime aspect of leftist histories is usually perceived as their negativity or darksidedness; as in most other negative approaches, this one also began with positive efforts, efforts to aid the disadvantaged. Yet, the negative attitude, not only toward corporate capitalism but also governments, leaders of most organizations, and technology, was dominant and only somewhat sharper than the generally rising pessimism that showed in many other histories.

Past synthetic histories have been rejected, first, because they lacked objectivity in pushing their own positive veiwpoints and, second, because those viewpoints were not now acceptable, especially after the rise of moral and new advocacy histories. Many synthetic generalizations were not merely challenged, they were eliminated—among them, heavy political favoritisms, the concept of progress, and acceptance of determinism. Much has been made of Herbert Butterfield's *Whig Interpretation of History.* As early as 1931 he made clear the unacceptability of pursuing history from a fixed political position or on the basis of evolutionary determinism and progress. Since then, the concept of progress has been much maligned, often on the incorrect assumption that progress must be accepted as deterministic. Progress, of course, is such a broad concept that it can be applied in widely differing ways, but its extensive elimination has contributed widely to the rise of pessimism.

Determinism has been even more thoroughly dismissed than progress, but its loss has not encouraged pessimism as much. Indeed, because Karl Marx was long regarded as a determinist, especially in his writings on technology, Marxists were under some pressure when the acceptability of determinism declined. There is now support, however, for rejecting Marx's determinism especially as applied to technology, thus giving that access to pessimism back to the Marxists.[26]

Support for pessimism or darkside history has had many sources, from both concentrated specializations and large-scale syntheses. Most localized specializations emphasize groups, regions, or topics that have received only limited academic study previously. The rise in studies of blacks, Indians, women, and labor have distinctly had this effect. Students who work in these fields are usually positive toward their topics, but this attitude leads them to become negative toward those groups with whom their subjects are in conflict. However, occasional scholars do sometimes begin their specialized study with a negative attitude toward it, an approach taken at least as rarely in the history of technology as in other fields. Indeed, historians of technology are less negative, even toward their related fields (such as government, business, science, and the elite) than are scholars in many other areas.

Large syntheses have become pessimistic as earlier themes and theses have been canceled out. The progressive sense of history and the concept of the United States as a uniquely different nation carried such historians as Charles Beard.[27] When they passed on, nothing of equivalent effectiveness replaced them, although for a time political democracy and "private acquisition" rose as strong themes. The political theme fell fastest, and John Diggins, although criticized for his view, concluded that after examining current and recent themes, "pessimism becomes obligatory."[28]

This rise in historians' negativity and in the understanding of the loss Americans generally experienced in their faith in national themes led to a series of "end" books. Among them were Henry May's *End of American Innocence,* Daniel Bell's *End of Ideology,* Andrew Hacker's *End of the American Era,* and David W. Noble's *End of American History.*[29] Although from different backgrounds, periods, and topics, they all stress negative change.

Two books on technology that reflect the darkside of the new left and Marxism, of specialized study, and of negative synthesis are David F. Noble's *America by Design* and *Forces of Production.* Although both have evoked some dissenting reactions, in common with current patterns they demonstrate good research and well-organized interpretation.[30] They represent the current problems of general history that are shared by the history of technology.

There are continuing moves in history to solve problems and to supply unfilled needs. Like changes in technology, they introduce disadvantages as well as advantages. Since the 1950s, increasingly amicable interactions have arisen with the humanities and the social sciences.[31] Some historians continue to be most comfortable within one and some in the other, but the major forward movements of recent years have been in or related to the social sciences. One of the most striking developments was the rise of quantitative methodology, which for a time produced sharp conflict. Today quantification has become very much a part of most history as have other social-science approaches. The literary and humanities approaches are still strong at points but have not grown as clearly. American studies, for example, which has represented primarily a history-literature approach, is not moving upward rapidly—although some history of technology publications of importance have emerged from this background.[32]

The increasing cooperation between related fields in general history is more than matched in certain specialized areas, notably the history of technology. This trend shows in the publications of many historians who come from or include science, economics, and social studies in their research. More dramatic is the combined annual meetings the Society for the

History of Technology holds with the History of Science Society, the Association of the Philosophy of Science, and the Society for the Study of the Science and Society.

The field of Science, Technology, and Society (STS) Studies that arose during the past fifteen years in a number of universities is composed of divergent fields of which history is only one element, although the history of technology has become particularly strong. The emphasis here on society may be responsible for leading STS studies to include negative dimensions so clearly. One commentator, Wilhelm Fudpucker, ties STS programs to movements originating in the 1960s, asserting that like black studies of that decade and women's studies of the 1970s, "STS studies grew out of, and what is more important, like them has abetted the diffusion of, a protest movement—in the case of STS studies, diffusing [and, perhaps, defusing as well?] the radical critique of the anti-technology and the environmental movements as deployed against government and industry in the 1960s and '70s."[33]

At any rate, STS and other combined studies, as well as history itself, including the history of technology, all involve increasing concern for social relationships. The rise of social history has abated, but that field remains of great importance not only on its own but through its integration into other fields. This development is also related to the social role of history. As noted earlier, history's social role is extremely important to all people, but, while recent developments have led certain groups and combinations to recognize this view, it is not usually accepted in the professional world.

Indeed, the attitudes that destroyed large-scale historical syntheses have usually been expressed with the feeling that the social intention of such generalizations is undesirable, that the celebration of group, religious, and national histories always contained even more undesirable positions, such as the promotion of the elite, of racism, of dubious governmental patterns, and of questionable business and economic patterns.

Other factors have also contributed to the decline of the planned and professionally approved social function of history; however, the decrease of large syntheses is the major loss. Henry Bausman has asserted correctly that "rightly or wrongly, the old histories served a clear, even though dubious social function since they described the rise and expected greatness of particular peoples and gave these peoples a sense of mission and purpose."[34] Historians may not accept the need for this role of history, but, if their work opposes or fails to contribute to the social need for positive synthesis, it will be obtained elsewhere. Indeed, movies, television, and novels already perform this function in a degree that most historians would like to see modified by the integration of better professional history.[35] They should at the same time begin to contribute to producing the positive historical syntheses that all people require.

This task, of course, is not easy because it seems so totally opposed to the ideal of objectivity. It is not, however, in any measure opposed to the goal of accuracy through the most precise and extensive historical research. The darkside historians with a negative purpose in mind have accomplished this; they have produced accurate and well-researched histories that are valuable for all historians. However, their purposes are not the primary need of the whole people who require history as a social memory that will permit them to move knowingly and successfully into the future.

It should be understood, in fact, that a large part of history is still motivated by positive objectives. Actually, much support for history is based on recognition of its role as a monument—a way of celebrating existing elements of the present by publicizing the history of their development. This is certainly the purpose of companies that have their own histories written, behind the massive number of historians hired by all the military services, and behind controlled national histories such as those of the Soviet Union and Japan. In another context, this same purpose is equally the overwhelming motivation behind black, women's, and labor histories, which are aimed at celebrating and aiding all three groups today by making available knowledge of their enormous and inadequately studied pasts. This purpose also had much to do with the rise of the history of technology. Not only engineers but journalists and general, science, and business historians of the early period usually anticipated that learning more about technology's past would celebrate its achievements and its major role today and tomorrow.

As noted, historians of technology have a responsibility for introducing technology into syntheses not only in their own field but also in general history. When this task is left to general historians, even those sensitive to the importance of technology, the results are too limited. For example, Howard Rock's *Artisans of the New Republic* is an excellent survey of New York artisans in the "age of Jefferson," and the best of his illustrations depict craftsmen at work. But none of the new technologies of the time, such as steamboating, are a part of his picture because they were not yet up to statistical numbering.[36] Similarly, William McNeill's *The Pursuit of Power: Technology, Armed Force, and Society since A.D. 1000* does discuss technology but only in terms of its social, industrial, and military connections; for the most part, technology itself is omitted. McNeill mentions almost no works by historians of technology; he does however refer to Melvin Kranzberg—but only in terms of his first book, *The Siege of Paris*, written before he entered the history of technology.[37]

However, McNeill's recognition of "mythistory" is a major contribution to understanding the need for purposeful syntheses designed to reach the people.[38] Historians of technology are well positioned to recognize the need for positive general syntheses and to integrate technology into them. They are in an even better position to increase effective history of technology syntheses.

This need must in no way reduce the sensitivity to wrongs, evils, and mistakes in and around technology. As in the effectiveness of the human memory, the recollection of one's past successes is not helpful unless the whole truth is included. For historians, this means that darksiders attacking the current and historical American world can still supply useful informa-tion—so long as they remain a minority. The only necessity is that the bulk of the history of technology remain positive, as it is now, and that most large-scale syntheses also be positive.

NOTES

1. Douglas Adair, *Fame and The Founding Fathers,* ed. Trevor Colbourn (New York: Norton, 1974), pp. 24–25.

2. Carl Becker, *Everyman His Own Historian* (New York: Appleton-Century-Crofts, 1935).

3. William H. McNeill, "History for Citizens," in *Teaching History Today,* ed. Henry S. Bausum (Washington, D.C.: American Historical Association, 1985), pp. 31–32.

4. David Thelen, "The Profession and *The Journal of American History,*" *Journal of American History* 73 (June 1986): 9.

5. Henry S. Bausum, "The Social Function of History," in Bausum, *Teaching History,* p. 41.

6. Brooke Hindle, "A Retrospective View of Science, Technology, and Material Culture in Early American History," *William and Mary Quarterly,* 3d ser. 41 (October 1984): 422–35.

7. Stanley I. Cutler and Stanley N. Katz, eds., *Reviews in American History* 10 (March 1982): vii.

8. Philip D. Curtin, "Depth, Span, and Relevance," *American Historical Review* 89 (February 1984): 1.

9. Roger Burlingame, *March of the Iron Men* (New York: Charles Scribner's Sons, 1938); idem, *Engines of Democracy* (New York: Charles Scribner's Sons, 1940).

10. Robert S. Woodbury, *History of the Gear-Cutting Machine* (Cambridge, Mass.: MIT Press, 1958); idem, *History of the Milling Machine* (Cambridge, Mass.: MIT Press, 1960). Abbott Payson Usher, *A History of Mechanical Inventions* (Cambridge, Mass.: Harvard University Press, 1954).

11. John M. Staudenmaier, S. J., *Technology's Storytellers: Reweaving the Human Fabric* (Cambridge, Mass.: MIT Press, 1985).

12. Melvin Kranzberg and Carroll W. Pursell, eds., *Technology in Western Civi-lization,* 2 vols. (New York: Oxford University Press, 1967).

13. Merritt Roe Smith, *Harpers Ferry Armory and the New Technology* (Ithaca, N.Y.: Cornell University Press, 1977).

14. Charles A. Reich, *The Greening of America* (New York: Random House, 1970), pp. 1, 5–7.

15. Ronald Berman, *America in the Sixties: An Intellectual History* (New York: Free Press, 1968), p. 147.

16. E. Bradford Burns, "Teaching History: A Changing Clientele and an Affirma-tion of Goals," in Bausum, *Teaching History,* p. 19.

17. Langdon Winner, *Autonomous Technology: Technics Out-of-Control as a Theme in Political Thought* (Cambridge, Mass.: MIT Press, 1977), p. x.

18. Ibid., p. 306.

19. Edward Hallett Carr, *What is History?* (New York: Alfred A. Knopf, 1967), p. 5.

20. Charles C. Gillispie, *The Edge of Objectivity* (Princeton, N.J.: Princeton University Press, 1960).

21. Robert Jones Shafer, ed., *A Guide to Historical Method* (Homewood, Ill.: Dorsey, 1969), p. 153.

22. See Carl N. Degler, "Remaking American History," *Journal of American History* 67 (June 1980): 7–25.

23. Michael Kammen, ed., *The Past before Us: Contemporary Historical Writing in the United States* (Ithaca: Cornell University Press, 1980), pp. 23–24.

24. Thomas Bender, "Wholes and Parts: The Need for Synthesis in American History," *Journal of American History* 73 (June 1986): 120, 121.

25. Irwin Unger, "The 'New Left' and American History: Some Recent Trends in American Historiography," *American Historical Review* 72 (July 1967): 1243, passim.

26. Donald MacKenzie, "Marx and the Machine," *Technology and Culture* 25 (July 1984): 473–502.

27. David W. Noble, *The End of American History: Democracy, Capitalism, and the Metaphor of Two Worlds in Anglo-American Historical Writing, 1880–1980* (Minneapolis: University of Minnesota Press, 1985), pp. 14, 141–43.

28. John Patrick Diggins, "Comrades and Citizens: New Mythologies in American Historiography; Comment," *American Historical Review* 90 (June 1985): 649.

29. Henry F. May, *The End of American Innocence* (New York: Alfred A. Knopf, 1959); Daniel Bell, *The End of Ideology* (New York: Free Press, 1960); Andrew Hacker, *The End of the American Era* (New York: Atheneum 1970); Noble, *End of American History.*

30. David F. Noble, *American by Design: Science, Technology, and the Rise of Corporate Capitalism* (New York: Alfred A. Knopf, 1977); idem, *Forces of Production: A Social History of Industrial Automation* (New York: Alfred A. Knopf, 1984).

31. Paul L. Ward, *Studying History: An Introduction to Methods and Structure,* 3d ed. (Washington, D.C.: American Historical Association, 1985).

32. E.g., John F. Kasson, *Civilizing the Machine: Technology and Republican Values in America, 1776–1900* (New York: Grossman, 1976).

33. Wilhelm E. Fudpucker, S.J., Letter to the Editor, in *Science, Technology, and Society: Curriculum Newsletter of the Lehigh University STS Program & Technology Studies Resource Center* 53 (April 1986): 12.

34. Bausum, "Social Function of History," p. 39.

35. On historical fiction, see Thomas Fleming, "Inventing our Probable Past," *New York Times Book Review,* 6 July 1986, pp. 1, 20–21.

36. Howard B. Rock, *Artisans of the New Republic: The Tradesmen of New York City in the Age of Jefferson* (New York: New York University Press, 1979).

37. William H. McNeill, *The Pursuit of Power: Technology, Armed Force, and Society since A.D. 1000* (Chicago: University of Chicago Press, 1982), passim, p. 253.

38. William H. McNeill, "Mythistory, or Truth, Myth, History, and Historians," *American Historical Review* 91 (February 1986): 1–10.

One Last Word—Technology and History: "Kranzberg's Laws"

MELVIN KRANZBERG

A few months ago I received a note from a longtime collaborator in building the Society for the History of Technology, Eugene S. Ferguson, in which he wrote, "Each of us has only one message to convey." Ferguson was being typically modest in referring to an article of his in a French journal[1] emphasizing the hands-on, design component of technical development, and he claimed that he had been making exactly the same point in his many other writings. True, but he has also given us many other messages over the years.

However, Ferguson's statement of "only one message" might indeed be true in my case. For I have been conveying basically the same message for over thirty years, namely, the significance in human affairs of the history of technology and the value of the contextual approach in understanding technical developments.

Because I have repeated that same message so often, utilizing various examples or stressing certain elements to accord with the interests of the different audiences I was attempting to reach, my thoughts have jelled into what have been called "Kranzberg's Laws." These are not laws in the sense of commandments but rather a series of truisms deriving from a longtime immersion in the study of the development of technology and its interactions with sociocultural change.

* * *

We historians tend to think of historical change in terms of cause and effect and of means and ends. Although it is not always easy to find causative elements and to distinguish ends from means in the interactions between technology and society, that has not kept scholars from trying to do so.

Indeed one of the intellectual clichés of our time, whose scholarly state-

Melvin Kranzberg originally presented this paper as his presidential address to the Society for the History of Technology on 19 October 1985, at the Henry Ford Museum in Dearborn, Michigan. It was subsequently published in *Technology and Culture* 28 (July 1986): 544–60, and is reprinted here by permission.

ment is embodied in the writings of Jacques Ellul and Langdon Winner, is that technology is pursued for its own sake and without regard to human need.[2] Technology, it is said, has become autonomous and has outrun human control; in a startling reversal, the machines have become the masters of man. Such arguments frequently result in the philosophical doctrine of technological determinism, namely, that technology is the prime factor in shaping our life-styles, values, institutions, and other elements of our society.

Not all scholars accept this version of technological omnipotence. Lynn White, jr., has said that a technical device "merely opens a door, it does not compel one to enter."[3] In this view, technology might be regarded as simply a means that humans are free to employ or not, as they see fit—and White recognizes that many nontechnical factors might affect that decision. Nevertheless, several questions do arise. True, one is not compelled to enter White's open door, but an open door is an invitation. Besides, who decides which doors to open—and, once one has entered the door, are not one's future directions guided by the contours of the corridor or chamber into which one has stepped? Equally important, once one has crossed the threshold, can one turn back?

Frankly, we historians do not know the answer to this question of technological determinism. Ours is a new discipline; we are still working on the problem, and we might never reach agreement on an answer—which means that it will provide employment for historians of technology for decades to come. Yet there are several things that we do know, and that I summarize under the label of Kranzberg's First Law.

Kranzberg's First Law reads as follows: Technology is neither good nor bad; nor is it neutral.

By that I mean that technology's interaction with the social ecology is such that technical developments frequently have environmental, social, and human consequences that go far beyond the immediate purposes of the technical devices and practices themselves, and the same technology can have quite different results when introduced into different contexts or under different circumstances.

Many of our technology-related problems arise because of the unforeseen consequences when apparently benign technologies are employed on a massive scale. Hence many technical applications that seemed a boon to mankind when first introduced became threats when their use became widespread. For example, DDT was employed to raise agricultural productivity and to eliminate disease-carrying pests. Then we discovered that DDT not only did that but also threatened ecological systems, including the food chain of birds, fishes, and eventually man. So the Western industrialized nations banned DDT. They could afford to do so, because their high technological level enabled them to use alternative means of pest control to achieve the same results at a slightly higher cost.

But India continued to employ DDT, despite the possibility of environmental damage, because it was not economically feasible to change to less persistent insecticides—and because, to India, the use of DDT in agriculture was secondary to its role in disease prevention. According to the World Health Organization, the use of DDT in the 1950s and 1960s in India cut the incidence of malaria in that country from 100 million cases a year to only 15,000, and the death toll from 750,000 to 1,500 a year. Is it surprising that the Indians viewed DDT differently from us, welcoming it rather than banning it? The point is that the same technology can answer questions differently, depending on the context into which it is introduced and the problem it is designed to solve.

Thus while some American scholars point to the dehumanizing character of work in a modern factory,[4] D. S. Naipaul, the great Indian author, assesses it differently from the standpoint of his culture, saying, "Indian poverty is more dehumanizing than any machine."[5] Hence in judging the efficacy of technological development, we historians must take cognizance of varying social contexts.

It is also imperative that we compare short-range and long-range impacts. In the nineteenth century, Romantic writers and social critics condemned industrial technology for the harsh conditions under which the mill workers and coal miners labored. Yet, according to Fernand Braudel, conditions on the medieval manor were even worse.[6] Certain economic historians have pointed out that, although the conditions of the early factory workers left much to be desired, in the long run the worker's living standards improved as industrialization brought forth a torrent of goods that were made available to an ever-wider public.[7] Of course, those long-run benefits were small comfort to those who suffered in the short run; yet it is the duty of the historian to show the differences between the immediate and long-range implications of technological developments.

Although our technological advances have yielded manifold benefits in increasing food supply, in providing a deluge of material goods, and in prolonging human life, people do not always appreciate technology's contributions to their lives and comfort. Nicholas Rescher, citing statistical data on the way people perceive their conditions, explains their dissatisfaction on the paradoxical ground that technical progress inflates their expectations faster than it can actually meet them.[8]

Of course, the public's perception of technological advances can change over time. A century ago, smoke from industrial smokestacks was regarded as a sign of a region's prosperity; only later was it recognized that the smoke was despoiling the environment. There were "technological fixes," of course. Thus, one of the aims of the Clean Air Act of 1972 was to prevent the harmful particulates emitted by smokestacks from falling on nearby communities. One way to do away with this problem was to build the smokestacks hundreds of feet high; then a few years later we discovered that the sulfur dioxide and other oxides, when sent high into the air, combined with water vapor to shower the earth with acid rain that has polluted lakes and caused forests to die hundreds of miles away.

Unforeseen "dis-benefits" can thus arise from presumably beneficent technologies. For example, although advances in medical technology and water and sewage treatment have freed millions of people from disease and plague and have lowered infant mortality, these have also brought the possibility of overcrowding the earth and producing, from other causes, human suffering on a vast scale. Similarly, nuclear technology offers the prospect of unlimited energy resources, but it has also brought the possibility of worldwide destruction.

That is why I think that my first law—Technology is neither good nor bad; nor is it neutral—should constantly remind us that it is the historian's duty to compare short-term versus long-term results, the utopian hopes versus the spotted actuality, the what-might-have-been against what actually happened, and the trade-offs among various "goods" and possible "bads." All of this can be done only by seeing how technology interacts in different ways with different values and institutions, indeed, with the entire sociocultural milieu.[9]

* * *

Whereas my first law stresses the interactions between technology and society, my second law starts with internalist elements in technology and then stretches to include many nontechnical factors. Kranzberg's Second Law can be simply stated: Invention is the mother of necessity.

Every technical innovation seems to require additional technical advances in order to make it fully effective. If one invents a lathe that can cut metal faster than existing machines, this necessitates improvements in the lubricating system to keep the mechanism running efficiently, improved grinding materials to stand up under the enhanced speed, and new means of taking away quickly the waste material from the item being turned.

Many major innovations have required further inventions to make them completely effective. Thus, Alexander Graham Bell's telephone spawned a variety of technical improvements, ranging from Edison's carbon-granule microphone to central-switching mechanisms. A variation on this same theme is described in Hugh Aitken's book on the origins of radio, in which he indicates the various innovative steps whereby the spark technology that produced radio waves was tuned into harmony (syntonized) with the receiver.[10] In more recent times, the design of a more powerful rocket, giving greater thrust, necessitates innovation in chemical engineering to produce the thrust, in materials to withstand the blast, in electronic control mechanisms, and the like.

A good case of invention mothering necessity can be seen in the landmark textile inventions of the eighteenth century. Kay's "flying shuttle" wove so quickly that it upset the usual ratio of four spinners to one weaver; either there had to be many more spinners or else spinning had to be similarly quickened by application of machinery. Thereupon Hargreaves, Cartwright, and Crompton improved the spinning process; then Cartwright set about further mechanizing the weaving operation in order to take full advantage of the now-abundant yarn produced by the new spinning machines.

Thomas P. Hughes would refer to the phenomenon that I have just described as a "reverse salient,"[11] but I prefer to call it a "technological imbalance," a situation in which an improvement in one machine upsets the previous balance and necessitates an effort to right the balance by means of a new innovation. No matter what one calls it, Hughes and I are talking about the same thing. Indeed, Hughes has gone further in discussing technological systems, for he shows how, as a system grows, it generates new properties and new problems, which in turn necessitate further changes.

The automobile is a prime example of how a successful technology requires auxiliary technologies to make it fully effective, for it brought whole new industries into being and turned existing industries in new directions by its need for rubber tires, petroleum products, and new tools and materials. Furthermore, large-scale use of the auto demanded a host of auxiliary technological activities—roads and highways, garages and parking lots, traffic signals, and parking meters.

While it might be said that each of these other developments occurred in response to a specific need, I claim that it was the original invention that mothered that necessity. If we look into the internal history of any mechanical device, we find that the basic invention required other innovative changes to make it fully effective and that the completed mechanism in turn necessitated changes in auxiliary and supporting technological systems, which, taken all together, brought many changes in economic and sociocultural patterns.

* * *

What I have just said is virtually a statement of my Third Law: Technology comes in packages, big and small.

The fact is that today's complex mechanisms usually involve several processes and components. Radar, for example, is a very complicated system, requiring specialized materials, power sources, and intricate devices to send out waves of the proper frequency, detect them when they bounce off an object, and then interpret them and place the results on a screen.

That might explain why so many different people have laid claim to inventing radar. Each is perfectly right in pointing out that he provided an element essential to the final product, but that final product is composed of many separate elements brought together in a system that could not function without every single one of the components. Thus radar is the product of a packaging process, bringing together elements of different technologies into a single device.

In his fascinating account of the development of mass production, David

A. Hounshell tells how many different experiments and techniques were employed in bringing Ford's assembly line into being.[12] Although many of the component elements were already in existence, Ford put these together into a comprehensive system—but not without having to develop additional technical capabilities, such as conveyor lines, to make the assembly process more effective.

My third law has been extended even further by Thomas P. Hughes's 1985 Dexter Prize-winning book *Networks of Power*. What I call "packages" Hughes more precisely and accurately calls "systems," which he defines as coherent structures composed of interacting, interconnected components.[13] When one component changes, other parts of the system must undergo transformations so that the system might continue to function. Hence the parts of a system cannot be viewed in isolation but must be studied in terms of their interrelations with the other parts.

Although Hughes concentrates on electric power systems, what he provides is a paradigm that is applicable to other systems—transportation, water supply, communications, and the like. And because entire systems interact with other systems, a system cannot be studied in isolation any more than can its component parts; hence one must also look at the interaction of these systems with the entire social, political, economic, and cultural environment. Hughes's book thus provides excellent case studies proving the validity of the first three of Kranzberg's Laws, and also of my fourth dictum.

* * *

Unfortunately, Kranzberg's Fourth Law cannot be stated so pithily as the first three. It reads as follows: Although technology might be a prime element in many public issues, nontechnical factors take precedence in technology-policy decisions.

Engineers claim that their solutions to technical problems are not based on mushy social considerations; instead, they boast that their decisions depend on the hard and measurable facts of technical efficiency, which they define in terms of input-output factors such as cost of resources, power, and labor. However, as Edward Constant has shown in studying the Kuhnian paradigm's applicability to technological developments, many complicated sociocultural factors, especially human elements, are involved, even in what might seem to be "purely technical" decisions.[14]

Besides, engineers do not always agree with one another; different fields of engineering might have different solutions to the same problem, and even within the same field they might disagree on what weight to assign to different trade-off factors. Indeed, as Stuart W. Leslie demonstrated in his Usher Prize article on "Charles F. Kettering and the Copper-cooled Engine,"[15] the most efficient device does not always win out even in what we might regard as a narrowly technical decision within a single industrial corporation. Although Kettering regarded his copper-cooled engine as a technical success, it never went into production. Why not? True, it had some technical "bugs," but these could not be successfully ironed out because of divisions between the research engineers and the production people—and because of the overall decision that the copper-cooled engine could not meet the corporate demand for immediate profit. So technical worth, or at least potential technical capability and efficiency, was not the decisive element in halting the copper-cooled engine.

In *Networks of Power* Hughes likewise demonstrates how nontechnical factors affected the efficient growth of electrical networks by comparing developments in Chicago, Berlin, and London. Private enterprise in Chicago, in the person of Samuel Insull, followed the path of the most efficient technology in seeking economies of scale. In Berlin and London, however, municipal governments were more concerned about their own authority than about technical efficiency, and political infighting meant that they lagged behind in developing the most economical power networks.

Technologically "sweet" solutions do not always triumph over political and social forces.[16] The debate a dozen years ago over the supersonic transport (SST) provides an example. Although the SST offered potential advantages, its development to the point where its feasibility and desirability could be properly determined was never allowed to take place. Economic factors might have underlain the decision to cut R&D funds for the SST, but the public decision seems also to have been based on a fear of the environmental hazards posed by the supersonic aircraft in commercial aviation.

Environmental concerns have indeed assumed a major place in public decisions regarding technical initiatives. These concerns are not groundless, for we have seen how certain technologies, employed without awareness of potential environmental effects, have boomeranged to present hazardous problems, despite their early beneficial effects. Many engineers believe that hysterical fear about technological development has so gripped our nation that people overlook the benefits provided by technology and concentrate on the dangers presented either by ill-conceived technological applications or by human error or oversight in technical operations. But who can blame the public, with Love Canal and Bhopal crowding the headlines?[17]

American politics has now become the battleground of special-interest groups, and few of these groups are willing to make the trade-offs required in many engineering decisions. In the case of potential environmental hazards, Daniel A. Koshland has stated that we can satisfy one or the other of the different groups, but only at a cost of something undesirable to the others.[18]

Especially politicized has been the question of nuclear power. The nuclear industry itself has been partly to blame for technological deficiencies, but the presumption of risk by the public, especially following the Three Mile Island and Chernobyl accidents, has affected the future of what was once regarded as a safe and inexhaustible source of power. The public fears possible catastrophic consequences from nuclear generators.

Yet the historical fact is that no one has been killed by commercial nuclear power accidents in this country. Contrast this with the 50,000 Americans killed each year by automobiles. But although antinuclear protestors picket nuclear power plants under construction, we never see any demonstrators bearing signs saying "Ban the Buick"!

Partly this is due to the public's perception of risk, rather than to the actual risks themselves.[19] People seek a zero-risk society. But as Aaron Wildavsky has so aptly put it, "No risk is the highest risk of all."[20] For it would not only petrify our technology but also stultify developmental growth in society along any lines.

Nevertheless, the fact that political considerations take precedence over purely technical considerations should not alarm us. In a democracy, that is as it should be. To deal with questions involving the interactions between technology and the ecology, both natural and social, we have devised new social instruments, such as "technology assessment," to evaluate the possible consequences of the applications of technologies before they are applied.

Of course, political considerations often continue to take precedence over the commonsensible results of comprehensive and impartial technological assessments. But at least there is the recognition that technological developments frequently have social, human, and environmental implications that go far beyond the intention of the original technology itself.

* * *

The fact that historians of technology must be aware of outside forces and factors affecting technology—from the human personality of the inventor to the larger social, economic, political, and cultural milieu—has led me to Kranzberg's Fifth Law: All history is relevant, but the history of technology is the most relevant.

In her presidential address to the Organization of American Historians several years ago, Gerda Lerner pointed out how history satisfies a variety of human needs, serving as a cultural tradition that gives us personal identity in the continuum of the past and future of the human enterprise.[21] Other apologists for the profession point out that history is one of the fundamental liberal arts and is essential as a key to an understanding of the future.

No one would quarrel with such worthy sentiments, but, to repeat questions raised by Eugene D. Genovese, "If so, how can we explain the dangerous decline in the teaching of history in our schools; the cynical taunt, 'What is history good for anyway?' "[22] Although historians might write loftily of the importance of historical understanding by civilized people and citizens, many of today's students simply do not see the relevance of history to the present or to their future. I suggest that this is because most history, as it is currently taught, ignores the technological element.

Two centuries ago the great German philosopher Immanuel Kant stated that the two great questions in life are (1) What can I know? and (2) What ought I do?

To answer Kant's first question, we can learn the history of the past. I look on history as a series of questions that we ask of the past in order to find out how our present world came into being. We call ours a "technological age." How did it get to be that way? That indeed is the major question that the history of technology attempts to answer. Our students know that they live in a technological age, but any history that ignores the technological factor in societal development does little to enable them to comprehend how their world came into being.

True, economic and business historians have perforce taken cognizance of those technological elements that had a mighty effect on their subject matter. Similarly, social historians of the *Annales* school have stressed how technology set the patterns of daily life for the vast majority of people throughout history, and Brooke Hindle, in a fine historiographical article, has indicated how some of our fellow historians have begun to see how technology impinges on their special fields of study.[23] But for the most part, social, political, and intellectual historians have been oblivious to the technological parameters of their own subjects.

Perhaps most guilty of neglecting technology are those concerned with the history of the arts and with the entire panoply of humanistic concerns. Indeed, in many cases they are disdainful of technology, regarding it as somehow opposed to the humanities. This might be because they regard technology solely in terms of mechanical devices and do not even begin to comprehend the complex nature of technological developments and their direct influences on the arts, to say nothing of their indirect influence on mankind's humanistic endeavors.

Yet anyone familiar with Cyril Stanley Smith's writings would be aware of the importance of the aesthetic impulse in technical accomplishments and of how these in turn amplified the materials and techniques available for artistic expression.[24] And any historian of art or of the Renaissance should perceive that such artistic masters as Leonardo and Michelangelo were also great engineers. That relationship continues today, as David Billington has shown in stressing the relationship of structural design and art.[25]

Today's technological age provides new technical capabilities to enlarge the horizons and means of expression for artists in every field. Advances in musical instruments have given larger scope to the imagination of composers and to musical interpretation by performers. The advent of photography, the phonograph, radio, movies, and television have not only given artists, composers, and dramatists new tools with which to exercise their vision and talents but have also enlarged the audience for music, drama, and the whole panoply of the arts. They also extend our audio and visual memory, enabling us to see, hear, and preserve the great works of the past and present.

In the field of learning and education, there is little point in belaboring the impact of writing tools, paper, the printing press, and, nowadays, radio and TV. But there is also an indirect influence of technology on education, one that makes it more possible than ever before in human history for larger numbers of people in the industrialized nations to take advantage of formal schooling.

Let me give a brief example drawn from American history. Thomas Jefferson was very proud of the educational system that he devised for the state of Virginia. But in his educational scheme, only a very small percentage could ever hope to ascend to the heights of a university education.

This is not because Jefferson was an elitist. Far from it! But the fact is that the agrarian technology of his time was not productive enough to allow large numbers of youth to participate in the educational process. From a very early age, children worked in the fields alongside their parents or, if they were town dwellers, were apprenticed to craftsmen. Only when great increases in agricultural and industrial productivity were made possible by revolutionary developments in technology did society acquire sufficient wealth to keep children out of the work force and enable them to attend school. As the nineteenth century progressed, first elementary education was made compulsory, then secondary education, and by the mid-twentieth century, America had grown so wealthy that it could afford a college education for all its citizens. True, some students drop out of high school before completing it, and not everyone going to college takes full advantage of the educational opportunities. But the fact is that the majority of Americans today have the equivalent education of the small segment of the upper-class elite in preindustrial society. In brief, technology has been a significant factor, not only in the pattern of our daily lives and in our workaday world, but also in democratizing education and the intellectual realm of the arts and humanities.

However, such vast generalizations might do little to convince the public of the wisdom of Stanley N. Katz's vision of scholars participating "in public discourse in order to recover the traditional role of the humanist as a public figure."[26] But the relevance of the history of technology to today's world can be spelled out in very specific terms. For example, because we live in a "global village," made so by technological developments, we are conscious of the need to transfer technological expertise to our less fortunate brethren in the less developed nations. And the history of technology has a great deal to say about the conditions, complexities, and problems of technology transfer.

Likewise, we are faced with public decisions regarding global strategy, environmental concerns, educational directions, and the ratio of resources to the world's burgeoning population. Technological history can cast light on many parameters of these very specific problems confronting us now and in the future—and that is why I say that the history of technology is more relevant than other histories.

One proof of this is that the outside world, especially the political community, is becoming increasingly cognizant of the contributions that historians of technology can make to public concerns. Whereas several decades ago historians were rarely called on to provide information to Congress on matters other than historical archives, memorials, and national celebrations, nowadays it is almost commonplace for historians of technology to testify before congressional committees dealing with scientific and technological expenditures, aerospace developments, transportation, water supplies, and other problems having a technological component. Congressmen obviously think that the information provided by historians of technology is relevant to coping with the problems of today and tomorrow.

Leaders in all fields are increasingly turning to historians of technology for expertise regarding the nature of the sociotechnical problems facing them. Let me give a few more specific examples. SHOT is an affiliate of the American Association for the Advancement of Science (AAAS), and there was a time when historians of technology appeared only on the program sessions of Section L of the AAAS, the History and Philosophy of Science. But historians of technology also have important things to say to a public larger than that composed of their historical colleagues. Hence it was a source of great personal pride to me—almost paternal pride—when, at the 1985 AAAS meeting, Carroll Pursell appeared on a program session with a congressman and a former assistant secretary of commerce; the program dealt with certain social and economic problems affecting the United States today, and Pursell's historical account of the technological parameters was truly germane to the thrust of the discussion. Similarly, at a recent conference, at my own Georgia Tech, on the problems expected to affect the workplace in the future, David Hounshell provided a meaningful technological historical context for a discussion that involved top labor leaders, political figures, and corporate executives. (I took family pride in that too!)

I regard this entrance of historians of technology into the public arena as empirical evidence of the true relevance of the history of technology to the worlds of today and tomorrow. To reiterate, all history is relevant, but the history of technology is most relevant. The rest of the world realizes that, and SHOT is working to make our historical colleagues from other fields recognize it too.

* * *

This brings me to my final law, Kranzberg's Sixth Law: Technology is a very human activity—and so is the history of technology.

Anthropologists and archeologists studying primate evolution tell us of the importance of purposive toolmaking in the formation of *Homo sapiens*. The physical development of our species is apparently inextricably bound up with cultural developments, so that technology is classed as one of the earliest and most basic of human cultural characteristics, one helping to develop language and abstract thinking. Or, to put it another way, man could not have become *Homo sapiens,* "man the thinker," had he not at the same time been *Homo faber,* "man the maker."

Man is a constituent element of the technical process. Machines are made and used by human beings. Behind every machine, I see a face—indeed, many faces: the engineer, the worker, the businessman or businesswoman, and, sometimes, the general and admiral. Furthermore, the function of the technology is its use by human beings—and sometimes, alas, its abuse and misuse.

To those who identify technology simply with the machines themselves, I use the computer as a metaphor to show the importance of the interaction of human and social factors with the technical elements—for computers require both the mechanical element, the "hardware," and the human element, the "software"; without the software, the machine is simply an inert device, but without the hardware, the software is meaningless. We need both, the human and the purely technical components, in order to make the computer a usable and useful piece of technology.

Those of you who were at our Silver Anniversary meeting in 1983 will recall that I told an anecdote, which I sometimes use to quiet my most voluble antitechnological humanistic colleagues. A lady came up to the great violinist Fritz Kreisler after a concert and gushed, "Maestro, your violin makes such beautiful music." Kreisler held his violin up to his ear and said, "I don't hear any music coming out of it."

You see, the instrument, the hardware, the violin itself, was of no use without the human element. But then again, without the instrument, Kreisler would not have been able to make music. The history of technology is the story of man and tool—hand and mind—working together. If the hardware is faulty or if the software is deficient, the sounds that emerge will be discordant; but when man and machine work together, they can make some beautiful music.

People sometimes speak of the "technological imperative," meaning that technology rules our lives. Indeed, they can point to many technical elements, such as the clock, that determine the character and pace of our daily existence. Likewise, the automobile determines where and how we Americans live, work, think, play, and pray.

But this does not necessarily mean that the "technological imperative," usually based on efficiency or economy, necessarily directs all our thoughts and actions. We can point to many technical devices that would make life simpler or easier for us but which our social values and human sensibilities simply reject. Thus, for example, Ruth Schwartz Cowan has shown in her Dexter Prize-winning book, *More Work for Mother,* how communal kitchens would be feasible and save the mother from much drudgery of food preparation. But our adherence to the concept of the home has made that technical solution unworkable; instead, we have turned to other technologies to ease the housework and cooking chores, albeit requiring more time and attention from mother.[27]

In other words, technological capabilities do not necessarily determine our actions. Indeed, how else can we explain why we have spent billions of dollars on nuclear power plants that we have had to abandon before they were completed? Obviously, other human factors proved more powerful than the combined technical and economic pressures.

Our reluctance to bow to the "technological imperative" is shown by the great efforts to make machines "user friendly"—and we are also embarking on the task of making humans "machine friendly" through educational programs in "technological literacy" and through the work of our SHOT special-interest groups to reach out to a wider public.

One final note on this point. Today's technology makes possible teleconferencing. Hence it would be cheaper to stay at home and have the papers and discussions of the SHOT meeting brought to us by telecommunication devices. But here we are, gathered together in Dearborn, Michigan, because we recognize that there is more to be derived from a SHOT meeting than the fine scholarly papers. There is the stimulation and camaraderie of being together and bouncing our ideas off one another in a face-to-face context. SHOT meetings are notable for their collegial atmosphere. Perhaps it is because we are still a relatively young discipline, so that the average age of historians of technology is probably younger than that of those in other, older fields. Or perhaps it is because we have very efficient program and local arrangements committees, which tend to our needs and provide the wherewithal for our conviviality.

All that is so, but I also believe that SHOT meetings are so friendly and wonderful because we are united in our pursuit of knowledge. Surely we sometimes disagree in our interpretations of the historical facts; we would be less than human if we did not, and we would not be doing our proper job as scholars if we accepted unquestioningly everything our colleagues said.

But more important, we are united in our concern to understand the past—and also look at the future. Remember that I pointed out earlier that Immanuel Kant said that the two great questions in life are, first, What can I know? and, second, What ought I do?

What we can know is how our present world came to be, and that requires a knowledge of the development of technology and of its interactions with culture and society—the very things for which we stand. But we also have a mission in relation to the second of Kant's great questions—What ought we do with our knowledge?—for we possess special capabilities because of our growing knowledge and understanding of technological developments and their varying interactions with the sociocultural milieu.

After all, we call ours a man-made world. And it is that, because mankind, with the aid of its technology, has fashioned our physical and social environment, our institutions, and other accoutrements of our society. But if ours is truly a man-made world, I claim that mankind can *re*-make it. And in that remaking process, the history of technology can play a very important role in enabling us to meet the challenges besetting mankind now and in the future.

That might seem a vain, utopian ideal. But historians of technology who have studied the great triumphs of the human mind and ingenuity embodied in mankind's technological accomplishments (and also mankind's failures) throughout the ages—such historians can indeed "dare to dream" of remaking ours into a better world.

NOTES

1. Eugene S. Ferguson, "La Fondation des machines modernes: des dessins," *Culture technique* 14 (June 1985): 182–207. *Culture technique* is the publication of the Centre de Recherche sur la Culture Technique, located in Paris under the direction of Jocelyn de Noblet. The June 1983 edition of *Culture technique,* dedicated to *Technology and Culture,* contained French translations of a number of articles from the SHOT journal.

2. Jacques Ellul, *The Technological Society* (New York: Alfred A. Knopf, 1964), and Langdon Winner, *Autonomous Technology: Technics Out-of-Control as a Theme in Political Thought* (Cambridge, Mass.: MIT Press, 1977).

3. Lynn White, jr., *Medieval Technology and Social Change* (Oxford: Oxford University Press, 1962), p. 28.

4. E.g., Christopher Lasch, *The Minimal Self: Psychic Survival in Troubled Times* (New York: W. W. Norton, 1984).

5. Quoted in Dennis H. Wrong, "The Case against Modernity," *New York Times Book Review,* 28 October 1984, p. 7.

6. Fernand Braudel, *The Structures of Everyday Life,* vol. 1 of *Civilization and Capitalism, 15th–18th Century* (New York: Harper & Row, 1981).

7. E.g., T. S. Ashton, *The Industrial Revolution, 1760–1830* (London: Oxford University Press, 1948), and David S. Landes, *The Unbound Prometheus: Technological Change and Industrial Development in Western Europe from 1750 to the Present* (Cambridge: Cambridge University Press, 1969).

8. Nicholas Rescher, *Unpopular Essays on Technological Progress* (Pittsburgh: University of Pittsburgh Press, 1980).

9. The "New Directions" program session at the 1985 SHOT annual meeting indicated that historians of technology are continuing to broaden their concerns and are indeed investigating new areas of the sociocultural context in relation to technological developments.

10. Hugh G. J. Aitken, *Syntony and Spark: The Origins of Radio* (New York: John Wiley, 1976).

11. Thomas P. Hughes, "Inventors: The Problems They Choose, the Ideas They Have, and the Inventions They Make," in *Technological Innovation: A Critical Review of Current Knowledge,* ed. Patrick Kelly and Melvin Kranzberg (San Francisco: San Francisco Press, 1978), pp. 166–82.

12. David A. Hounshell, *From the American System to Mass Production 1800–1932: The Development of Manufacturing Technology in the United States* (Baltimore: Johns Hopkins University Press, 1984), chap. 6.

13. Thomas P. Hughes, *Networks of Power: Electrification in Western Society, 1880–1930* (Baltimore: Johns Hopkins University Press, 1983), p. ix.

14. Edward W. Constant, *The Origins of the Turbojet Revolution* (Baltimore: Johns Hopkins University Press, 1980). This book was awarded the Dexter Prize by SHOT in 1982.

15. Stuart W. Leslie, "Charles F. Kettering and the Copper-cooled Engine," *Technology and Culture* 20 (October 1979): 752–76.

16. Eugene B. Skolnikoff states, "Technology alters the physical reality, but is not the key determinant of the political changes that ensue," in *The International Imperatives of Technology: Technological Development and the International Political System* (Berkeley, Calif.: University of California Institute of International Studies, n.d.), p. 2.

17. Speaking of the Bhopal tragedy, President John S. Morris of Union College has said: "Methyl isocyanate makes it possible to grow good crops and feed millions of people, but it also involves risks. And analyzing risks is not a simple matter" (*New York Times*, 14 April 1985).

18. Daniel A. Koshland, "The Undesirability Principle," *Science* 229 (5 July 1985): 9.

19. See Dorothy Nelkin, ed., *Controversy: The Politics of Ethical Decisions* (Santa Monica, Calif.: Sage, 1984).

20. Aaron Wildavsky, "No Risk Is the Highest Risk of All," *American Scientist* 67 (1979): 32–37.

21. Gerda Lerner, "The Necessity of History and the Professional Historian," *Journal of American History* 69 (June 1982): 7–20.

22. Eugene D. Genovese, "To Celebrate a Life—Biography as History," *Humanities* 1 (January–February 1980): 6. An analysis of today's low state of the history profession is to be found in Richard O. Curry and Lawrence D. Goodheart, "Encounters with Clio: The Evolution of Modern American Historical Writing," *OAH Newsletter* 12 (May 1984): 28–32.

23. Brooke Hindle, " 'The Exhilaration of Early American Technology': A New Look," in *The History of American Technology: Exhilaration or Discontent?* ed. David A. Hounshell (Greenville, Del.: Hagley Museum and Library, 1984).

24. See especially Cyril Stanley Smith's Usher Prize article, "Art, Technology, and Science: Notes on Their Historical Interaction," *Technology and Culture* 11 (October 1970): 493–549.

25. See David Billington's Dexter Prize-winning book, *Robert Maillart's Bridges: The Art of Engineering* (Princeton: Princeton University Press, 1979), and "Bridges and the New Art of Structural Engineering," *American Scientist* 72 (January–February 1984): 22–31.

26. Stanley N. Katz, "The Scholar and the Public," *Humanities* 6 (June 1985): 14–15.

27. Ruth S. Cowan, *More Work for Mother: The Ironies of Household Technology from the Open Hearth to the Microwave* (New York: Basic Books, 1983), chap. 5.

A Select Bibliography of the Publications of Melvin Kranzberg

Melvin Kranzberg's scholarly output of books, articles, and essays has been pro-digious—more than 175 in all. Here, we have not listed every item but rather have limited the selection to substantial publications closely related to the focus of this volume and to the history of technology and technology studies more generally, including such topics as technology assessment and technology transfer. We have omitted references to newspaper and encyclopedia articles and to many brief essays. In those instances where articles have been reprinted numerous times, we have indicated only the primary publication reference. The resulting list, covering some three decades, still numbers more than eighty entries. This corpus of work reflects Mel's longstanding commitment to a historically informed study of technology in its societal context.

BOOKS

Technology in Western Civilization (coeditor, Carroll W. Pursell). 2 vols. New York: Oxford University Press, 1967. Vol. 2 translated into Japanese by T. Kobayashi (2 vols.), Tokyo: Tokyo Keizai Shimposha, 1976. Vol. 1 translated into Spanish by Esteve Riambau i Sauri (2 vols.). Barcelona: Editorial Gustavo Gili, 1981.

Technology and Culture: An Anthology (coeditor, William Davenport). New York: Schocken Books, 1972; paperbound, Meridian Series, New York: New American Library, 1975. Translated into Arabic by Abdel Megeed Nassar, Cairo: Sigel al-Arab Publications, 1976. Translated into Spanish by Esteve Riambau i Sauri, Barcelona: Editorial Gustavo Gili, 1980.

By the Sweat of Thy Brow: Work in the Western World (coauthor Joseph Gies). New York: G. P. Putnam's, 1975; paperbound, Capricorn edition, 1976; reprinted hardbound, Westport, Conn.: Greenwood Press, 1986. Translated into Italian (*Breve storia del lavoro*) by G. Canavese and U. Livini. Milan: Arnoldo Mondadori Editore, 1976. Portions reprinted in *Work in Modern Society: A Sociology Reader,* edited by Lauri Perman, pp. 2–6, 40–46. Dubuque, Iowa: Kendall/Hunt, 1986.

"Confrontation with Technology," *Lex et Scientia: The International Journal of Law and Science* 12 (January–September 1976): 1–74. Special issue of journal devoted to Kranzberg's Mellon Lectures at Lehigh University.

Technological Innovation: A Critical Review of Current Knowledge (coeditor, Patrick Kelly). San Francisco: San Francisco Press, 1978.

Energy and the Way We Live (coeditor, Timothy A. Hall). San Francisco: Boyd & Fraser, 1980.

Ethics in an Age of Pervasive Technology (editor). Boulder, Colo.: Westview Press, 1980.

Bridge to the Future: A Centennial Celebration of the Brooklyn Bridge (coedited with Margaret Latimer and Brooke Hindle). Annals of the New York Academy of Sciences, vol. 424. New York: New York Academy of Sciences, 1984.

American Scientist (guest editor). Sigma Xi Centennial Issue, 74 (September–October 1986): 449–568.

Technological Education/Technological Style (editor). San Francisco: San Francisco Press, 1986.

ARTICLES AND ESSAYS

"At the Start." *Technology and Culture* 1 (Winter 1959): 1–10.

"Charles Singer and *A History of Technology.*" *Technology and Culture* 1 (Fall 1960): 299–302.

"Criteria for an Industrial Revolution." *Archives Internationales d'Histoire des Sciences* 105 (1960): 256–62. Also published in *Actes du IXe Congrès International d'Histoire des Sciences,* Barcelona: Associatión Para la Historia de la Ciencia Española; Paris: Hermann, 1960, pp. 212–20.

"The Newest History: Science and Technology." *Science* 136 (11 May 1962): 463–68. Reprinted in *Air Force and Space Digest* 45 (August 1962): 53–55.

"The Technical Act: Commentary." Proceedings of the Encyclopaedia Britannica Conference on the Technological Order, *Technology and Culture* 3 (Fall 1962): 519–23. Reprinted in *The Technological Order,* edited by Carl Stover. Detroit: Wayne State University Press, 1963.

"Technology and Human Values." *Virginia Quarterly Review* 40 (Autumn 1964): 578–92. Reprinted in *Technology and Human Affairs,* edited by Larry Hickman and Azizah Al-Hibri, pp. 393–99. St Louis: C. V. Mosby, 1979. Also reprinted in *Philosophy, Technology and Human Affairs,* edited by Larry Hickman, pp. 232–40. College Station, Tex.: Ibis Press, 1985.

"The Technological Revolution and Social Reform." In *New Perspectives on World History,* edited by Shirley H. Engle, pp. 33–62. 34th Yearbook of the National Council for the Social Sciences. Washington, D.C.: National Council for the Social Sciences, 1964.

"Accidental Enlightenment." *Virginia Quarterly Review* 41 (Autumn 1965): 635–38.

"The Unity of Science-Technology." *American Scientist* 55 (March 1967): 48–66. Reprinted in *Aspects of Science-Technology,* edited by Masao Watanabe, pp. 57–79. Tokyo: Kenkyusha, Ltd., 1973.

"Man and Megamachine." *Virginia Quarterly Review* 43 (Autumn 1967): 686–93.

"Technology is Important—Really It Is." *Industrial Arts Education* 27 (September–October 1967): 29–32.

"The Transfer of Technology: Imitative or Innovative?" In *The Transfer of Technology to Developing Nations,* edited by Daniel L. Spencer and Alexander Woroniak, pp. 30–34. New York: Praeger, 1967.

"The Spectrum of Science-Technology." *Journal of the Scientific Laboratories* 48 (December 1967): 47–58.

"The Disunity of Science-Technology." *American Scientist* 56 (Spring 1968): 21–34.

"Problems of Innovation in Technological Change." In *Actes du XIe Congrès International d'Histoire des Sciences, 1965* 6 (Warsaw: Polish Academy of Sciences, 1968), pp. 11–15.

"A Call to Action." Review of *The American Challenge*, by Jean-Jacques Servan-Schreiber. *Virginia Quarterly Review* 44 (Autumn 1968): 658–64.

"Technology: A Force for Social Change." *American Vocational Journal* 45 (February 1970): 30–33.

"An Interdisciplinary Team Approach from the Engineering Point of View." In *Industrialization and Development*, edited by H. E. Hoelscher and M. C. Hawk, pp. 338–46. San Francisco: San Francisco Press, 1969.

"The Multi-Dimensional Society." Review of *Marshall, Marx and Modern Times*, by Clark Kerr. *Virginia Quarterly Review* 46 (Spring 1970): 325–45.

"Drucker as Historian of Technological Change." In *Peter Drucker: Contributions to Business Enterprise*, edited by Tony H. Bonaparte and John E. Flaherty, pp. 337–61. New York: New York University Press, 1970.

"The New Role of the Humanities and Social Sciences." *Educational Technology* (June 1971): 31–33.

"Science-Technology and Warfare: Action, Reaction, and Interaction in the Post–World War II Era." In *Science, Technology, and Warfare: Proceedings of the Third Military History Symposium*, U.S. Air Force Academy, 8–9 May 1969, edited by Monte D. Wright and Lawrence D. Paszek, pp. 123–70. Washington, D.C.: U. S. Government Printing Office, 1971.

"The Human Use of Technology: Historical Perspective." *Tennessee Engineer* 4 (1970–71): 10–24.

"The Transfer of Technology to Colonial America." *Actes du XIIe Congrès International d'Histoire des Sciences* 11, Paris, 1968. (Paris: Albert Blanchard, 1971), pp. 81–85.

"Science, Technology, and the Unity of Mankind." In *History and The Idea of Mankind*, edited by W. Warren Wagar, pp. 133–58. Albuquerque: University of New Mexico Press, 1971.

"Technology and Humanism." In *Technology, Power, and Social Change*, edited by C. A. Thrall and J. M. Starr, pp. 112–22. Lexington, Mass.: D. C. Heath, 1972.

"Can Technological Progress Continue to Provide for the Future?" *The Greek Review of Social Research* 18 (1973): 158–65. Also published in *The Economic Growth Controversy*, edited by A. Weintraub, E. Schwartz, and J. Aronson. White Plains, N.Y.: International Arts and Sciences Press, 1973.

"Transferring a Dynamic Technology During Socio-Political Change: The American Experience." In *L'Acquisition des Techniques par les Pays Non-Initiateurs*, pp. 255–76. Paris: Centre National de la Recherche Scientifique, 1973.

"Environmental Problems and Technology." *Environment and Culture* (English translation of Japanese title) 12 (October 1974): 2–9.

"The Ecological Implications of Technology Transfer." In *XIVth International Congress of the History of Science, Japan*, pp. 273–94. Tokyo: Science Council of Japan, 1975.

"Development of Social and Communal Responsibility." In *A Technology Assessment Primer*, edited by Leon Kirchmayer, Harold Linstone, and William Marsh, pp. 12–14. New York: Institute of Electrical and Electronics Engineers, 1975.

"Are We Running Out of Time?" *Epsilon Pi Tau Journal* 11 (April 1975): 3–10.

"The Acquisition of Technology by Developing Countries." *The Indian and Eastern Engineer* (May 1975): 219–24. Also published in *Landmark* 2 (April 1975): 1–3.

"Historical Perspective on Industrialization." *Iowa State Journal of Research* 50 (August 1975): 81–87.

"Confrontation: Technology and the Social Environment." In *The Technological Catch and Society,* edited by Iraj Zandi, pp. 59–83. Philadelphia: University of Pennsylvania, 1975.

"The Cultural and Operational Distinctions between Science and Technology." In *AAAS Interdisciplinary Workshop on the Interrelationships Between Science and Technology, and Ethics and Values,* edited by William Blanpied and Wendy Weisman-Dermer, pp. 29–36. Washington, D.C.: American Association for the Advancement of Science, 1975.

"The Future Imperative of Technology." In *Vanderbilt University Symposium on Technology and Public Policy,* edited by Howard L. Hartman, pp. 17–21. Nashville, Tenn.: Vanderbilt University Press, 1976.

"Technology and Human Labor," "Science and Technology: Their Interdependence and Inter-Impact on the Contemporary State," and "The Influence of Cosmonautics on Different Spheres of the Activity of Man." In *Acta,* Special Issue 8 of *Czechoslovak Studies in the History of Science,* edited by Lobos Novy, pp. 47–52, 209–21, 331–37. Prague: Czechoslovak Academy of Sciences, 1976.

"The Social Ecology of Technology." In *Technology and Society: The Oak Ridge Bicentennial Lectures,* pp. 1–19. Oak Ridge, Tenn.: Oak Ridge National Laboratory, 1977. Reprinted in *Engineering and Humanities,* edited by James H. Schaub and Sheila K. Dickison, pp. 489–503. New York: Wiley-Interscience, 1982.

"Technology the Liberator." In *Technology at the Turning Point,* edited by William B. Pickett, pp. 35–47. San Francisco: San Francisco Press, 1977.

"Why Technology Assessment?" In *Impact of Solutions: Assessing Impacts of Northeast Issues,* pp. 4–13. Burlington: Vermont State Planning Office, 1977.

"Designing the Engineering Future." In *Proceedings: Conference on University Education for Technology and Public Policy,* edited by Eric B. Hartman and Robert P. Morgan, pp. 81–86. St. Louis, Mo.: Department of Technology and Human Affairs, Washington University, 1977.

"From Carpetbag to Carpet Mill: Technology in the New South." In *The Southern Mystique: Technology and Human Values in a Changing Region,* edited by W. David Lewis and B. Eugene Griessman, pp. 33–46. University: University of Alabama Press, 1977.

"Technology and Democratization of American Society." In *America's World Role for the Next Twenty-Five Years,* edited by Philip M. Chen, pp. 153–72. Taiwan: Tamkang College, 1977.

"The Science-Technology Complex." *transaction/SOCIETY* 15 (January/February 1978): 54–55.

"Technology and Human Values." *Dialogue* 11 (1978): 11–19.

"History and Diffusion of Innovations: A Postcript." In *The Diffusion of Innovations,* edited by M. Radnor, chap. 13. Evanston, Ill.: Northwestern University Center for the Interdisciplinary Study of Science and Technology, 1978.

"Technology Assessment in America." In *Technology and Its Impact on Society,* edited by Sigvard Strandh, pp. 235–54. Stockholm: Tekniska Museet, 1979. Translated into French as "L'évaluation technologique american," *Culture Technique* 10 (June 1983): 237–47.

"Materials in History and Society" (Cyril Stanley Smith, coauthor). *Materials Science and Engineering* 37 (January 1979): 1–39. Reprinted in *Materials Science and Engineering: Its Evolution, Practice, and Prospects,* edited by Morris Cohen, pp. 1–39. Lausanne, Switzerland: Elsevier Sequoia, 1981.

"The Human Origins of Technology." *Georgia Journal of Sciences* 37 (January 1979): 43–51.

"Introduction: Trends in the History and Philosophy of Technology." In *The History and Philosophy of Technology,* edited by George Bugliarello and Dean B. Doner, pp. xiii–xxxi. Urbana: University of Illinois Press, 1979.

"Technology the Civilizer." *Iowa State Journal of Research* 54 (November 1979): 163–73.

"Technology Transfer to Developing Nations." *Topic* 115 (n.d.), pp. 2–6.

"Communication of Scientific and Technical Information: Implications for Federal Policies and Research." In *The Five-Year Outlook: Problems, Opportunities, and Constraints in Science and Technology* 2 (Washington, D.C.: National Science Foundation, 1980), pp. 508–19.

"Technology: The Half-Full Cup." *Alternative Futures* 3 (Spring 1980): 5–18.

"Prospects for Change." In *Societal Risk Assessment: How Safe is Safe Enough?,* edited by Richard C. Schwing and Walter A. Albers, Jr., pp. 319–33. New York: Plenum, 1980.

"Scientific Research and Technical Innovation." *National Forum* 61 (Winter 1981): 27–28.

"Passing the Baton." *Technology and Culture* 22 (October 1981): 695–99.

"The History of Technology in the United States" (in Chinese). *Journal of Dialectics of Nature* 4 (1982): 24–34. Published by Academia Sinica, Beijing.

"The Industrialization of Western Society: 1860–1914." In *Science, Technology, and Society in the Time of Alfred Nobel,* edited by Carl Gustaf Bernhard, Elisabeth Crawford, and Per Sorbom, pp. 209–30. Oxford: Pergamon Press, 1982.

"Quirks and Jerks of Editing *Technology and Culture.*" In *The History and Sociology of Technology,* Special Publication in Anthropology and History 3, edited by Donald Hoke, pp. 3–15. Milwaukee: Milwaukee Public Museum, 1982.

"The History of Technology in the United States." *Historia Scientiarum* 23 (1982): 1–14. Published by the History of Science Society of Japan.

"Antitechnology Rampant." *Hastings Center Report* 12 (August 1982): 41–43.

"The Anonymity of the History of Technology," and "Concluding Presentation." In Sources for the History of Technology: National Comparisons, *Acta Historiae rerum Naturalium necnon Technicarum,* pp. xxiv–xxxviii, 437–56. Special Issue 17, Czechoslovak Academy of Sciences. Prague: Czechoslovak Academy of Sciences, 1982.

"Energy Past and Future." In *Energy for the Future: A Call for Leadership,* edited by A. R. Buhl, pp. 15–20. Proceedings of WATTec Energy Conference. Knoxville, Tenn.: Technology for Energy Corporation, 1983.

"Le processus d'innovation: Un model écologique." *Culture Technique* 10 (June 1983): 263–77.

"L'evolution technologique américaine." *Culture Technique* 10 (June 1983): 237–49.

"Let's Not Get Wrought Up about It." Comments on J. M. Staudenmaier, "What SHOT Hath Wrought and What SHOT Hath Not: Reflections on Twenty-five Years of the History of Technology." *Technology and Culture* 25 (October 1984): 735–49.

"The Wedding of Science and Technology: A Very Modern Marriage." *In Technology and Science: Important Distinctions for Liberal Arts Colleges,* edited by John Nicholas Burnet, pp. 27–37. Davidson, N.C.: Davidson College, 1984.

"Where Were We Going? Where Have We Gone?" *Progress* 35 (September–October

1984): 10–15. Publication of Keynote Speech at 50th Anniversary of Museum of Science and Industry, Chicago, Ill.

"Technological Revolutions." *National Forum* 64 (Summer 1984): 6–10. Reprinted in Social Issues Resources Series, *Technology* 2 (1985), article 29.

"Technology's the Answer—But That's Not the Question." In *Human Values in a High-Tech Society,* edited by Patricia N. Williamsen, pp. 2–12. Columbus: Ohio Humanities Council, 1985. Also printed in *Urban Resources* 3 (Fall 1985): 3–4, 62.

"The Information Age: Evolution or Revolution?" In *Information Technologies and Social Transformation,* edited by Bruce Guile, pp. 35–53. Washington, D.C.: National Academy Press, 1985. Reprinted in *Economic Impact* 55 (1986): 67–73.

"The Technical Elements in International Technology Transfer: Historical Perspectives." In *The Political Economy of International Technology Transfer,* edited by John R. McIntyre and Daniel S. Papp, pp. 31–45. Westport, Conn.: Quorum Books, Greenwood Press, 1986.

"Machine-Made America: Technology and the Democratization of American Society." *Polhem* (Journal of the Swedish National Committee for the History of Technology) 4 (1986): 233–51.

"Technology and History: Kranzberg's Laws." *Technology and Culture* 27 (July 1986): 544–60.

Selective Bibliography of Publications Cited

The following bibliography contains a selective listing of major secondary books and articles dealing with the history of technology that have been cited by the authors in this volume. It is included both to provide an easy reference to the works cited and to suggest something of the breadth and coverage of the field. It is by no means intended to be comprehensive, but it does provide a useful overview of the field as it has evolved to its present state.

Readers wishing to pursue specific topics in the history of technology further will find a number of bibliographic sources helpful. Among the most useful are Eugene S. Ferguson's detailed *Bibliography of the History of Technology* (Cambridge, Mass.: SHOT and MIT Press, 1968), and Marc Rothenberg's selective overview, *The History of Science and Technology in the United States: A Critical and Selective Bibliography* (New York: Garland, 1982). This latter volume is part of a very useful extended series of annotated bibliographies being published by Garland on the histories of specific sciences and technologies. Volumes on chemical technology, civil engineering, mining and metallurgy, and the technologies of the Greeks and Romans and of the medieval period have appeared as well. The reader is also referred to the Society for the History of Technology's "Current Bibliography in the History of Technology," published in the April issue of *Technology and Culture,* for annually updated bibliographic information.

Likewise helpful are two particularly good bibliographies in the history of science: George Sarton's *A Guide to the History of Science* (New York: Ronald, 1952) and Francois Russo's *Histoire des Sciences et des Techniques: Bibliographie,* 2d ed. (Paris: Hermann, 1969), both of which contain references in the history of technology as well as science. In this vein the annual *Critical Bibliography of the History of Science* contained in the journal *Isis,* and Magda Withrow's *ISIS Cumulative Bibliography. A Bibliography of the History of Science* (London and Bronx, N.Y.: Mansell, 1971–) are also very helpful.

Finally, in the broader field of science and technology studies one should consult Stephen H. Cutcliffe, Judith Mistichelli, and Christine Roysdon's *Technology and Values in American Civilization: A Guide to Information Sources* (Detroit: Gale Research, 1980) and Paul Durbin's *A Guide to the Culture of Science, Technology and Medicine,* 2d ed. (New York: Free Press, 1984), both of which have useful bibliographies and, in the case of the latter, essays that address major themes and issues of the fields.

Adams, Edward Dean. *Niagara Power: History of the Niagara Falls Power Company, 1886–1918.* 2 vols. Niagara Falls, N.Y.: Niagara Falls Power Co., 1927.

Agassi, Joseph. *Technology: Philosophical and Social Aspects.* Dordrecht: Reidel, 1985.

Agricola, Georgius. *De Re Metallica*. Translated by Herbert Clark Hoover and Lou Henry Hoover. New York: Dover, 1950.

Aitken, Hugh G. J. *Syntony and Spark: The Origins of Radio*. New York: John Wiley, 1976.

Althin, Torsten K. W. *C. E. Johansson, 1864–1943*. Translated by Cyril Marshall. Stockholm: Privately printed, 1948.

Anderson, Oscar F. *Refrigeration in America: A History of a New Technology and Its Impact*. Princeton: Princeton University Press, 1953.

Bathe, Greville, and Dorothy Bathe. *Oliver Evans*. Philadelphia: The Historical Society of Pennsylvania, 1935.

Belfield, Robert. "The Niagara System: The Evolution of an Electric Power Complex at Niagara Falls." *Proceedings of the IEEE* 64 (September 1976): 1344–50.

Billington, David P. *Robert Maillart's Bridges, the Art of Engineering*. Princeton: Princeton University Press, 1979.

———. *The Tower and the Bridge*. New York: Basic Books, 1983.

Boorstin, Daniel. *The Americans*. 3 vols. New York: Vintage Books, 1958–73.

Borgmann, Albert. *Technology and the Character of Contemporary Life: A Philosophical Inquiry*. Chicago: University of Chicago Press, 1985.

Braun, Hans-Joachim. "The National Association of German-American Technologists and Technology Transfer between Germany and the United States, 1884–1930." *History of Technology* 8 (1983): 15–35.

Brittain, James E. "The Introduction of the Loading Coil: George A. Campbell and Michael I. Pupin." *Technology and Culture* 11 (January 1970): 36–57.

Bryant, Lynwood. "The Development of the Diesel Engine." *Technology and Culture* 17 (July 1986): 432–46.

———. "The Origin of the Four-Stroke Cycle." *Technology and Culture* 8 (April 1967): 178–98.

———. "The Role of Thermodynamics in the Evolution of Heat Engines." *Technology and Culture* 14 (April 1973): 152–65.

———. "Rudolph Diesel and His Rational Engine." *Scientific American* 221 (August 1969): 108–18.

———. "The Silent Otto." *Technology and Culture* 7 (Spring 1966): 184–200.

Buchanan, R. A. "The Promethean Revolution: Science, Technology, and History." *History of Technology* 1 (1976): 73–83.

———. "Technology and History." *Social Studies of Science* 5 (1975): 489–99.

Bugliarello, George, and Dean B. Doner, eds. *The History and Philosophy of Technology*. Urbana: University of Illinois Press, 1979.

Bunge, Mario. *Treatise on Basic Philosophy*, Vol. 7, Pt. 2. Dordrecht: Reidel, 1985.

Burke, John G. "The Complex Nature of Explanations in the Historiography of Technology." *Technology and Culture* 11 (January 1970): 22–26.

———. "Wood Pulp, Water Pollution, and Advertising." *Technology and Culture* 20 (January 1979): 175–95.

Burlingame, Roger. *Engines of Democracy*. New York: Charles Scribner's Sons, 1940.

———. *The March of the Iron Men*. New York: Charles Scribner's Sons, 1938.

Burn, D. L. "The Genesis of American Engineering Competition, 1850–1870." *Economic History* 2 (1931): 292–311.

Callahan, Daniel. *The Tyranny of Survival: And Other Pathologies of Civilized Life*. New York: Macmillan, 1973.

Cardwell, Donald S. L. "The Academic Study of the History of Technology." *History of Science* 7 (1968): 112–24.

——. *From Watt to Clausius, the Rise of Thermodynamics in the Early Industrial Age*. Ithaca, N.Y.: Cornell University Press, 1971.

——. "Some Factors in the Early Development of the Concepts of Power, Work, and Energy." *British Journal for the History of Science* 3 (1966–67): 209–24.

——. *Turning Points in Western Technology*. New York: Science History Press, 1972.

Channell, David F. "The Harmony of Theory and Practice: The Engineering Science of W. J. M. Rankine." *Technology and Culture* 23 (January 1982): 39–52.

Chapelle, Howard I. *The History of the American Sailing Navy*. New York: Norton, 1949.

Chapkis, Wendy, and Cynthia Enloe, eds. *Of Common Cloth: Women in the Global Textile Industry*. Amsterdam and Washington, D.C.: Transnational Institute, 1983.

Childe, V. Gordon. *Man Makes Himself.* New York: Mentor, 1951.

Cipolla, Carlo. *Guns, Sails, and Empires: Technological Innovation and the Early Phase of European Expansion, 1400–1700*. New York: Minerva, 1965.

Coates, Vary T., and Bernard Finn. *A Retrospective Technology Assessment: Submarine Telegraphy—The Transatlantic Cable of 1866*. San Francisco: San Francisco Press, 1979.

Cochran, Thomas C. *Frontiers of Change*. New York: Oxford University Press, 1981.

Condit, Carl W. "Sullivan's Skyscrapers as Expressions of Nineteenth Century Technology." *Technology and Culture* 1 (Winter 1959): 78–93.

Constant, Edward W. *The Origins of the Turbojet Revolution*. Baltimore: Johns Hopkins University Press, 1980.

Cooper, Carolyn C. "The Portsmouth System of Manufacture." *Technology and Culture* 25 (April 1984): 182–225.

——. "The Production Line at Portsmouth Block Mill." *Industrial Archaeology Review* 6 (Winter 1981–82): 28–44.

Cowan, Ruth Schwartz. "A Case Study of Technological and Social Change: The Washing Machine and the Working Wife." In *Clio's Consciousness Raised: New Perspectives on the History of Women*, edited by Mary Hartman and Lois W. Banner, pp. 245–53. New York: Harper and Row, 1974.

——. "From Virginia Dare to Virginia Slims: Women and Technology in American Life." *Technology and Culture* 20 (January 1979): 51–63.

——. "The 'Industrial Revolution' in the Home: Household Technology and Social Change in the 20th Century." *Technology and Culture* 17 (January 1976): 1–23.

——. *More Work for Mother: The Ironies of Household Technology from the Open Hearth to the Microwave*. New York: Basic Books, 1983.

Cummins, C. Lyle, Jr. *Internal Fire*. Lake Oswego, Ore.: Carnot Press, 1976.

Cutcliffe, Stephen. "Retrospective Technology Assessment." *STS Newsletter* (Lehigh University) 18 (June 1980): 7–12.

Czitrom, Daniel J. *Media and the American Mind: From Morse to McLuhan*. Chapel Hill: University of North Carolina Press, 1982.

Daniels, George H. "The Big Questions in the History of Technology." *Technology and Culture* 11 (1970): 1–21. Also in *The State of American History*, edited by Herbert J. Bass, pp. 197–219. Chicago: Quadrangle, 1970.

Darby, H. C. *The Draining of the Fens.* Cambridge: Cambridge University Press, 1940.

Davies, Robert Bruce. *Peacefully Working to Conquer the World: Singer Sewing Machines in Foreign Markets, 1854–1920.* New York: Arno, 1976.

Davies, Margery W. *Woman's Place is at the Typewriter: Office Work and Office Workers, 1870–1930.* Philadelphia: Temple University Press, 1982.

De Solla Price, Derek J. "Is Technology Historically Independent of Science? A Study in Statistical Historiography." *Technology and Culture* 6 (Fall 1965): 553–68.

Dessauer, Freidrich. *Philosophie der Technik: Das Problem der Realisierung.* Bonn: F. Cohen, 1927.

———. *Streit um die Technik.* Frankfurt: J. Knecht, 1956.

Dickinson, H. W., A. A. and Gomme. " 'Netherlands' Contribution to Great Britain's Engineering and Technology to the Year 1700." *Archives Internationales d'Histoire des Sciences* No. 3 (April 1950): 356–77.

———. "Some British Contributions to Continental Technology, 1600–1850." *Archives Internationales d'Histoire des Sciences* No. 16 (July 1951): 706–22.

Dretske, Fred I. *Knowledge and the Flow of Information.* Cambridge, Mass.: MIT Press, 1981.

Drucker, Peter F. "Modern Technology and Ancient Jobs." *Technology and Culture* 4 (Summer 1963): 277–81.

Durbin, Paul T., ed. *A Guide to the Culture of Science, Technology and Medicine.* New York: Free Press, 1980, 1984.

———. *Research in Philosophy and Technology.* Greenwich, Conn.: JAI Press, 1978–.

Durbin, Paul T., and Friedrich Rapp, eds. *Philosophy and Technology.* Boston Studies in the Philosophy of Science, vol. 80. Dordrecht: Reidel, 1983.

Ellul, Jacques. *The Technological Society,* 1954. Translated by John Wilkinson. New York: Alfred A. Knopf, 1964.

———. *The Technological System,* 1977. Translated by Joachim Neugroschel. New York: Continuum, 1980.

Ferguson, Eugene S. "The American-ness of American Technology." *Technology and Culture* 20 (January 1979): 3–24.

———. "The Mind's Eye: Nonverbal Thought in Technology." *Science* 197 (26 August 1977): 827–36.

———. "On the Origin and Development of American Mechanical 'Know-How'." *Midcontinent American Studies Journal* 3 (Fall 1962): 3–16.

———. "Towards a Discipline of the History of Technology." *Technology and Culture* 15 (January 1974): 13–30.

Finch, James Kip. *A History of the School of Engineering, Columbia University.* New York: Columbia University Press, 1954.

Fisher, Marvin W. "The Iconology of Industrialism, 1830–1860." *American Quarterly* 13 (Fall 1961): 347–64.

———. *Workshops in the Wilderness: European Response to American Industrialization, 1830–1860.* New York: Oxford University Press, 1967.

Flink, James J. "The Three Stages of American Automobile Consciousness." *American Quarterly* 24 (October 1972): 451–73.

———. *The Car Culture.* Cambridge, Mass.: MIT Press, 1975.

Florman, Samuel C. *The Existential Pleasures of Engineering.* New York: St. Martin's, 1976.

Foley, Vernard. "Leonardo's Contributions to Theoretical Mechanics." *Scientific American* 225 (September 1986): 108–13.

Foley, Vernard, and Werner Soedel. "Ancient Oared Warships." *Scientific American* 244 (April 1981): 148–63.

Gendron, Bernard. *Technology and the Human Condition.* New York: St. Martin's, 1977.

Giedion, Sigfried. *Mechanization Takes Command: A Contribution to Anonymous History.* New York: Oxford University Press, 1948.

———. *Space, Time, and Architecture: The Growth of a New Tradition.* Cambridge, Mass.: Harvard University Press, 1941.

Gillispie, Charles C. *Lazare Carnot, Savant.* Princeton: Princeton University Press, 1971.

Goldman, Steven L. "The *Technē* of Philosophy and the Philosophy of Technology." In *Research in Philosophy & Technology,* Vol. 7, edited by Paul T. Durbin, pp. 115–44. Greenwich, Conn.: JAI Press, 1984.

Gutman, Herbert. *Work, Culture and Society in Industrializing America.* New York: Vintage, 1966.

Habakkuk, H. J. *American and British Technology in the Nineteenth Century.* Cambridge: Cambridge University Press, 1962.

Hacker, Barton C. "Greek Catapults and Catapult Technology: Science, Technology and War in the Ancient World." *Technology and Culture* 9 (January 1968): 34–50.

Hacker, Sally L. "The Culture of Engineering: Woman, Workplace and Machine." *Women's Studies International Quarterly* 4 (1981): 341–53.

———. "Farming out the Home: Women and Agribusiness," *The Second Wave* 5 (Spring/Summer 1977): 38–49.

———. "Sex Stratification, Technology and Organizational Change: A Longitudinal Case Study of AT&T." *Social Problems* 26 (June 1979): 539–57.

Halahan, B. C. "Chiddingford Glass and Its Makers in the Middle Ages." *Transactions of the Newcomen Society* 8 (1926): 188–95.

Hall, A. Rupert. "On Knowing and Knowing How to. . . ." *History of Technology* 3 (1978): 91–102.

Harris, John R., and T. C. Barker. *A Merseyside Town in the Industrial Revolution.* Liverpool: University Press, 1954.

Haynes, William. *The American Chemicals Industry.* 6 vols. New York: Van Nostrand, 1945.

Headrick, Daniel R. *The Tools of Empire: Technology and European Imperialism in the Nineteenth Century.* New York: Oxford University Press, 1981.

Heidegger, Martin. *The Question Concerning Technology and Other Essays.* New York: Harper and Row, 1977.

Heilbroner, Robert. "Do Machines Make History?" *Technology and Culture* 8 (July 1967): 335–45.

Hindle, Brooke. "British v. French Influence on Technology in the Early United States." In *Actes du XIe Congrès International d'Histoire des Sciences, 1965* 6 (Warsaw: Polish Academy of Sciences, 1968), pp. 49–53.

———. *Emulation and Invention.* New York: New York University Press, 1981.

————. "The Exhilaration of Early American Technology." In *Technology in Early America*. Chapel Hill: University of North Carolina Press, 1966.

————. "How Much is a Piece of the True Cross Worth?" In *Material Culture and the Study of American Life*, edited by Ian M. G. Quimby, pp. 5–20. New York: W. W. Norton, 1978.

————. "A Retrospective View of Science, Technology, and Material Culture in Early American History." *William and Mary Quarterly* 41 (October 1984): 422–35.

Hindle, Brooke, and Stephen Lubar. *Engines of Change: The American Industrial Revolution 1790–1860*. Washington, D. C.: Smithsonian Institution, 1986.

Hoffecker, Carol E. *Corporate Capital: Wilmington in the Twentieth Century*. Philadelphia: Temple University Press, 1983.

Hooven, Frederick J. "The Wright Brothers' Flight-Control System." *Scientific American* 239 (November 1978): 167–82.

Hounshell, David A. *From the American System to Mass Production, 1800–1932*. Baltimore: Johns Hopkins University Press, 1984.

————, ed. *The History of American Technology: Exhilaration or Discontent?* Hagley Papers. Greenville, Del.: Hagley Museum and Library, 1984.

————. "Public Relations or Public Understanding? The American Industry Series in *Scientific American*." *Technology and Culture* 21 (October 1980): 589–93.

Hounshell, David A., and John K. Smith. *Science and Corporate Strategy: DuPont R&D 1902–1980*. New York: Cambridge University Press, 1988.

Hughes, Thomas P. *Elmer Sperry: Inventor and Engineer*. Baltimore: Johns Hopkins University Press, 1971.

————. "Emerging Themes in the History of Technology." *Technology and Culture* 20 (October 1979): 697–711.

————. *Networks of Power: Electrification in Western Society, 1880–1930*. Baltimore: Johns Hopkins University Press, 1983.

————. "The Order of the Technological World." *History of Technology* 5 (1980): 1–16.

————. "We Get the Technology We Deserve." *American Heritage* 36 (October–November 1985): 65–79.

Hulten, Pontus, K. G., ed. *The Machine as Seen at the End of the Machine Age*. New York: Museum of Modern Art, 1968.

Huning, Alois. "Philosophy of Technology and the Verein Deutscher Ingenieure." In *Research in Philosophy and Technology*, Vol. 2, edited by Paul T. Durbin, pp. 265–71. Greenwich, Conn.: JAI Press, 1979.

Hunter, Louis C. *Steamboats on the Western Rivers*. Cambridge, Mass.: Harvard University Press, 1949.

Ihde, Don. *Technics and Praxis*. Boston Studies in the Philosophy of Science, vol. 24. Dordrecht: Reidel, 1979.

Jenkins, C. Rhys. "Notes on the Early History of Steel Making in England." *Transactions of the Newcomen Society* 3 (1922): 16–23.

————. "Observations on the Rise and Progress of Manufacturing Industry in England." *Transactions of the Newcomen Society* 7 (1926): 1–15.

————. "The Rise and Fall of the Sussex Iron Industry." *Transactions of the Newcomen Society* 1 (1920): 16–33.

Jenkins, Reese. *Images and Enterprise: Technology and the American Photographic Industry, 1839–1925*. Baltimore: Johns Hopkins University Press, 1975.

Jeremy, David J. *Transatlantic Industrial Revolution: The Diffusion of Textile Technologies between Britain and America, 1790–1830.* Cambridge, Mass.: MIT Press, 1981.

Johnson, William A., trans. *Christopher Polhem: The Father of Swedish Technology.* Hartford, Conn.: Trinity College, 1963.

Jonas, Hans. *The Imperative of Responsibility: In Search of an Ethics for the Technological Age.* Chicago: University of Chicago Press, 1984; German original, 1979.

Jones, Daniel P. "From Military to Civilian Technology: The Introduction of Tear Gas for Civil Riot Control." *Technology and Culture* 19 (April 1978): 151–68.

Kaempffert, Waldemar, ed. *A Popular History of American Invention,* 2 vols. New York: Charles Scribner's Sons, 1924.

Kapp, Ernst. *Grundlinien einer Philosophie der Technik: Zur Entstehungsgeschichte der Cultur aus neuen Gesichtspunkt,* 1877. Reprinted with new introduction. Dusseldorf: Janssen, 1978.

Kasson, John F. *Civilizing the Machine: Technology and Republican Values in America, 1776–1900.* New York: Grossman, 1976.

Keller, Alexander G. "Has Science Created Technology?" *Minerva* 22 (1984): 160–82.

———. "The Missing Years of Jacques Besson, Inventor of Machines, Teacher of Mathematics, Distiller of Oils, and Huguenot Pastor." *Technology and Culture* 14 (January 1973): 28–39.

Keller, Evelyn Fox. *Reflections on Gender and Science.* New Haven, Conn.: Yale University Press, 1985.

Kline, Ronald R. "Scientific Electrotechnology: The Case of the Induction Motor." *Technology and Culture* 28 (April 1987): 283–313.

Kouwenhoven, John A. *Made in America: The Arts and Modern Civilization.* Garden City, N.Y.: Doubleday, 1948.

Kranakis, Eda Fowlks. "The French Connection: Giffard's Injector and the Nature of Heat." *Technology and Culture* 23 (January 1982): 3–38.

Kranzberg, Melvin. "At the Start." *Technology and Culture* 1 (Winter 1959): 1–10.

———. "Passing the Baton." *Technology and Culture* 22 (October 1981): 695–99.

———. "Technology and History: 'Kranzberg's Laws'." *Technology and Culture* 27 (July 1986): 544–60.

Kranzberg, Melvin, and Carroll W. Pursell, eds. *Technology in Western Civilization,* 2 vols. New York: Oxford University Press, 1967.

Krohn, Wolfgang, et al., eds. *The Dynamics of Science and Technology: Social Values, Technical Norms and Scientific Criteria in the Development of Knowledge.* Dordrecht: D. Reidel, 1978.

Lambie, Joseph, ed. *Architects and Craftsmen in History.* Tubingen: Mohr, 1956.

Layton, Edwin T., Jr. "European Origins of the American Engineering Style of the Nineteenth Century." In *Scientific Colonialism: A Cross-Cultural Comparison,* edited by Nathan Reingold and Marc Rothenberg, pp. 151–66. Washington, D.C.: Smithsonian Institution, 1987.

———. *The Revolt of the Engineers: Social Responsibility and the American Engineering Profession.* Cleveland: Press of Case Western Reserve University, 1971; Baltimore: Johns Hopkins University Press, 1986.

———. "Scientific Technology, 1845–1900: The Hydraulic Turbine and the Origins of American Industrial Research." *Technology and Culture* 20 (January 1979): 64–89.

———. "Technology as Knowledge." *Technology and Culture* 15 (January 1974): 31–41.

———, ed. *Technology and Social Change in America.* New York: Harper and Row, 1973.

Lenk, Hans. *Zur Sozialphilosophie der Technik.* Frankfurt: Suhrhamp, 1982.

Licht, Walter. *Working for the Railroad: The Organization of Work in the Nineteenth Century.* Princeton: Princeton University Press, 1983.

Lindqvist, Svante. *The Teaching of History of Technology in USA—A Critical Survey in 1978.* Stockholm Papers in History and Philosophy of Technology. Stockholm: Royal Institute of Technology Library, 1981.

McCullough, David. *The Great Bridge.* New York: Simon and Shuster, 1972.

McDougall, Walter A. *. . . the Heavens and the Earth: A Political History of the Space Age.* New York: Basic Books, 1985.

McGaw, Judith A. *Most Wonderful Machine: Mechanization and Social Change in Berkshire Paper Making, 1801–1885.* Princeton: Princeton University Press, 1987.

———. "Women and the History of American Technology." *Signs: Journal of Women in Culture and Society* 7 (Summer 1982): 798–828.

McHugh, Jeanne. *Alexander Holley and the Makers of Steel.* Baltimore: Johns Hopkins University Press, 1980.

MacKenzie, Donald. "Marx and the Machine." *Technology and Culture* 25 (July 1984): 473–502.

McMahon, Michal. "An American Courtship: Psychologists and Advertising Theory in the Progressive Era." *American Studies* 13 (1972): 5–18.

McNeill, William H. *The Pursuit of Power: Technology, Armed Forces, and Society since A.D. 1000.* Chicago: University of Chicago Press, 1982.

Man, Science, Technology: A Marxist Analysis of the Scientific and Technological Revolution. Moscow and Prague: Academia, 1973.

Marchand, Roland. *Advertising the American Dream, Making for Modernity, 1920–1940.* Berkeley: University of California Press, 1985.

Marcuse, Herbert. *One-Dimensional Man: Studies in the Ideology of Advanced Industrial Society.* Boston: Beacon, 1964.

Mark, Robert, and William W. Clark. "Gothic Structural Experimentation." *Scientific American* 251 (November 1984): 176–84.

Mark, Robert, and Ronald S. Jonash. "Wind Loading on Gothic Structure." *Journal of the Society of Architectural Historians* 29 (October 1970): 222–30.

Marx, Leo. *The Machine in the Garden: Technology and the Pastoral Ideal in America.* New York: Oxford University Press, 1964.

Mayr, Otto. "Yankee Practice and Engineering Theory: Charles T. Porter and the Dynamics of the High-Speed Steam Engine." *Technology and Culture* 16 (October 1975): 570–602.

Mazlish, Bruce, ed. *The Railroad and the Space Program.* Cambridge, Mass.: MIT Press, 1965.

Mazuzan, George T. "Atomic Power Safety: The Case of the Power Reactor Development Company Fast Breeder, 1955–1956." *Technology and Culture* 23 (July 1982): 341–71.

———. " 'Very Risky Business': A Power Reactor for New York City." *Technology and Culture* 27 (April 1986): 262–84.

Melosi, Martin V., ed. *Pollution and Reform in American Cities, 1870–1930*. Austin: University of Texas Press, 1980.

Merchant, Carolyn. *The Death of Nature: Women, Ecology, and the Scientific Revolution*. New York: Harper and Row, 1980.

Merton, R. K. *Science, Technology and Society in 17th-century England (Osiris* 4 [1938]: 360–632). New York: H. Fertig, 1970.

Meyer, Stephen. *The Five Dollar Day: Labor Management and Social Control in the Ford Motor Company, 1908–1921*. Albany: State University of New York Press, 1981.

Miller, Perry. "The Responsibility of Mind in a Civilization of Machines." *American Scholar* 31 (Winter 1961–62): 51–69.

Mitcham, Carl. "What is the Philosophy of Technology?" *International Philosophical Quarterly* 25 (March 1985): 73–88.

Mitcham, Carl, and Alois Huning, eds. *Philosophy and Technology II: Information Technology and Computers in Theory and Practice*. Boston Studies in the Philosophy of Science, vol. 90. Dordrecht: Reidel, 1986.

Mitcham, Carl, and Robert Mackey. *Bibliography of the Philosophy of Technology*. Chicago: University of Chicago Press, 1973.

————, eds. *Philosophy and Technology: Readings in the Philosophical Problems of Technology*. 1972, 2d ed. New York: Free Press, 1983.

Multhauf, Robert P. "The Scientist and the 'Improver' of Technology." *Technology and Culture* 1 (Winter 1959): 38–47.

————. "Some Observations on the State of the History of Technology." *Technology and Culture* 15 (January 1974): 1–12.

Mumford, Lewis. *The Myth of the Machine*, vol. 1: *Technics and Human Development*. New York: Harcourt Brace Jovanovich, 1967. vol. 2: *The Pentagon of Power*. New York: Harcourt Brace Jovanovich, 1970.

————. *Technics and Civilization*. New York: Harcourt Brace, 1934.

Nef, J. U. *Industry and Government in France and England, 1540–1640*. Ithaca, N.Y.: Cornell University Press, 1957.

Nevins, Allan, and Frank Ernest Hill. *Ford*, 3 vols. New York: Scribners, 1954–1963.

Noble, David F. *America by Design: Science, Technology, and the Rise of Corporate Capitalism*. New York: Alfred A. Knopf, 1977.

————. *Forces of Production: A Social History of Industrial Automation*. New York: Alfred A. Knopf, 1984.

————. "Technology's Politics: Present Tense Technology." *Democracy* 3 (Spring 1983): 8–24.

Pool, Ithiel de Sola. *The Social Impact of the Telephone*. Cambridge, Mass.: MIT Press, 1977.

Pope, Daniel. *The Making of Modern Advertising*. New York: Basic Books, 1983.

Post, Robert C. *Physics, Patents, and Politics. A Biography of Charles Grafton Page*. New York: Science History Publications, 1976.

————. "In Praise of Top Fuelers." *American Heritage of Invention and Technology* 1 (Spring 1986): 58–63.

Pupin, Michael. *From Immigrant to Inventor*. New York: Charles Scribner's Sons, 1925.

Pursell, Carroll W., Jr. "History of Technology." In *A Guide to the Culture of Science,*

Technology, and Medicine, edited by Paul T. Durbin, pp. 70–120. New York: Free Press, 1980, 1984.

———. "The History of Technology and the Study of Material Culture." *American Quarterly* 35 (1983): 304–15.

———. "Testing a Carriage: The 'American Industry' Series of *Scientific American.*" *Technology and Culture* 17 (January 1976): 82–92.

———, ed. *Technology in America: A History of Individuals and Ideas.* Cambridge, Mass.: MIT Press, 1981.

Rae, John B. "The 'Know-How' Tradition in American History." *Technology and Culture* 1 (Spring 1960): 139–50.

Randall, Adrian J. "The Philosophy of Luddism: The Case of the West of England Woolen Workers, ca. 1790–1809." *Technology and Culture* 27 (January 1986): 1–17.

Rapp, Friedrich. *Analytical Philosophy of Technology,* translated from the German original (1978) by Stanley R. Carpenter and Theodore Langenbruch. Boston Studies in the Philosophy of Science, vol. 63. Dordrecht: Reidel, 1981.

———. "Philosophy of Technology." In *Contemporary Philosophy: A New Survey,* vol. 2: *Philosophy of Science,* edited by G. Floistad, pp. 361–412. The Hague: Nijhoff, 1982.

———, ed. *Contributions to a Philosophy of Technology: The Structure of Thinking in the Technological Sciences.* Dordrecht: Reidel, 1974.

Rapp, Friedrich, and Paul T. Durbin, eds. *Technikphilosophie in der Diskussion.* Braunschweig and Wiesbaden: Vieweg, 1982.

Ravetz, Alison. "Modern Technology and an Ancient Occupation: Housework in Present Day Society." *Technology and Culture* 6 (Spring 1965): 256–60.

Read, Thomas T. *The Development of Mineral Industry Education in the United States.* New York: American Institute of Mining and Metallurgical Engineers, 1941.

Reich, Leonard. "Industrial Research and the Pursuit of Corporate Security: The Early Years of Bell Labs." *Business History Review* 54 (Winter 1980): 504–29.

———. *The Making of American Industrial Research: Science and Business at GE and Bell, 1876–1926.* Cambridge and New York: Cambridge University Press, 1985.

Reti, Ladislao, ed. *The Unknown Leonardo.* New York: McGraw-Hill, 1974.

Reti, Ladislao, and Bern Dibner, eds. *Leonardo da Vinci, Technologist.* Norwalk, Conn.: Burndy Library, 1969.

Reynolds, Terry S. "Medieval Roots of the Industrial Revolution." *Scientific American* 251 (July 1984): 122–30.

———. *Stronger than a Thousand Men, A History of the Vertical Water Wheel.* Baltimore: Johns Hopkins University Press, 1983.

Richter, Jean Paul, ed. *The Notebooks of Leonardo da Vinci, Compiled and Edited from the Original Manuscripts,* 2 vols. New York: Dover, 1970.

Rock, Howard B. *Artisans of the New Republic: The Tradesmen of New York City in the Age of Jefferson.* New York: New York University Press, 1979.

Roe, Joseph W. *English and American Tool Builders.* New Haven, Conn.: Yale University Press, 1916.

Rogin, Leo. *The Introduction of Farm Machinery in Its Relation to the Productivity of Labor in the United States during the Nineteenth Century.* Berkeley: University of California Press, 1931.

Rosenberg, Nathan. *Technology and American Economic Growth.* New York: Harper and Row, 1972.

———, ed. *The American System of Manufactures: The Report of the Committee on the Machinery of the United States 1855 and the Special Reports of George Wallis and Joseph Whitworth 1854.* Edinburgh: University Press, 1969.

Rothschild, Joan, ed. *Machina ex Dea: Feminist Perspectives on Technology.* New York: Pergamon Press, 1983.

———. *Teaching Technology from a Feminist Perspective: A Practical Guide.* New York: Pergamon, 1987.

Rothschild, Emma. *Paradise Lost: The Decline of the Auto-Industrial Age.* New York: Random House, 1973.

Rouse, Hunter, and Simon Ince. *History of Hydraulics.* New York: Dover, 1957.

Rürup, Reinhard. "Historians and Modern Technology." *Technology and Culture* 15 (April 1974): 161–93.

Schofield, Robert E. *The Lunar Society of Birmingham: A Social History of Provincial Science and Industry in Eighteenth-Century England.* Oxford: Clarendon, 1963.

Sawyer, John E. "The Social Basis of the American System of Manufacturing." *Journal of Economic History* 14 (Winter 1954): 361–79.

Scientific Technology and Social Change: Readings from Scientific American. San Francisco: Freeman, 1974.

Segal, Howard P. "Assessing Retrospective Technology Assessment." *Technology in Society* 4 (1982): 231–46.

Shammas, Carole. "Consumer Behavior in Colonial America." *Social Science History* 6 (Winter 1982): 67–88.

———. "The Domestic Environment in Early Modern England and America." *Journal of Social History* 14 (Fall 1980): 3–24.

———. "How Efficient Was Early America?" *Journal of Interdisciplinary History* 13 (Autumn 1982): 247–72.

Shrader-Frechette, Kristin S. *Nuclear Power and Public Policy: The Social and Ethical Problems of Fission Technology,* 2d ed. Dordrecht: Reidel, 1983.

———. *Science Policy, Ethics, and Economic Methodology: Some Problems of Technology Assessment and Environmental Impact Analysis.* Dordrecht: Reidel, 1984.

Simon, Herbert. *Sciences of the Artificial.* Cambridge, Mass.: MIT Press, 1969.

Sinclair, Bruce. *Philadelphia's Philosopher Mechanics: A History of the Franklin Institute, 1824–1865.* Baltimore: Johns Hopkins University Press, 1974.

Sinclair, Bruce. "The Promise of the Future: Technical Education." In *Nineteenth Century American Science,* edited by George H. Daniels, pp. 249–72. Evanston, Ill.: Northwestern University Press, 1972.

Singer, Charles. "How 'A History of Technology' Came Into Being." *Technology and Culture* 1 (Fall 1960): 302–11.

Singer, Charles, A. R. Hall, and Trevor I. Williams, et al., eds. *History of Technology,* 8 vols. London: Oxford University Press, 1954–84.

Smith, Cyril S. "Art, Technology and Science: Notes on Their Historical Interaction." *Technology and Culture* 11 (October 1970): 493–549.

Smith, Merritt Roe. *Harpers Ferry Armory and the New Technology: The Challenge of Change.* Ithaca, N.Y.: Cornell University Press, 1977.

————. "Social Processes and Technological Change." *Reviews in American History* 13 (June 1985): 157–66.

————, ed. *Military Enterprise and Technological Change: Perspectives on the American Experience.* Cambridge, Mass.: MIT Press, 1985.

Smith, Norman A. F. "The Origin of the Water Turbine and the Invention of its Name." *History of Technology* 2 (1977): 215–59.

Soedel, Werner, and Vernard Foley. "Ancient Catapults." *Scientific American* 240 (March 1979): 150–60.

Spence, Clark C. *Mining Engineers and the American West: The Lace Boot Brigade, 1849–1933.* New Haven: Yale University Press, 1970.

Stapleton, Darwin H. *Accounts of European Science, Technology, and Medicine Written by American Travelers Abroad, 1735–1860, in The Collection of the American Philosophical Society.* Philadelphia: American Philosophical Society Library, 1985.

Stapleton, Darwin H., and David A. Hounshell. "The Discipline of the History of American Technology: An Exchange." *Journal of American History* 68 (1982): 897–902.

Staudenmaier, John M., S. J. *Technology's Storytellers: Reweaving the Human Fabric.* Cambridge, Mass.: MIT Press, 1985.

————. "What SHOT Hath Wrought and What SHOT Hath Not: Reflections on Twenty-five Years of the History of Technology." *Technology and Culture* 25 (October 1984): 707–30.

Steinman, David B. *The Builders of the Bridge.* New York: Harcourt, Brace & Co., 1945.

Stephens, Carlene. *Inventing Standard Time.* Washington, D.C.: Smithsonian Institution, 1983.

Strasser, Susan. *Never Done: A History of American Housework.* New York: Pantheon, 1982.

Struik, Dirk J. *Yankee Science in the Making.* Boston: Little, Brown and Company 1948.

Tarr, Joel A., ed. "The City and Technology." *Journal of Urban History* 5 (May 1979): 275–406.

————. *Retrospective Technology Assessment.* San Francisco: San Francisco Press, 1976.

Tarr, Joel A., and Francis C. McMichael. *Retrospective Assessment of Wastewater Technology in the United States: 1800–1872.* Pittsburgh: Carnegie Mellon University, 1977.

Tarr, Joel, et al. "Water and Wastes: A Retrospective Assessment of Wastewater Technology in the United States, 1800–1932." *Technology and Culture* 25 (April 1984): 226–63.

TeBrake, William. "Air Pollution and Fuel Crises in Preindustrial London, 1250–1650." *Technology and Culture* 16 (July 1975): 337–59.

Thomas, Donald E., Jr. *Diesel: Technology and Society in Industrial Germany.* University: University of Alabama Press, 1987.

Thompson, Robert L. *Wiring a Continent: The History of the Telegraph Industry in the United States, 1832–1866.* Princeton: Princeton University Press, 1947.

Trachtenberg, Alan. *Brooklyn Bridge: Fact and Symbol.* New York: Oxford University Press, 1965.

Trescott, Martha Moore. *The Rise of the American Electrochemical Industry, 1880–1910. Studies in the American Technological Environment.* Westport, Conn.: Greenwood, 1981.

————, ed. *Dynamos and Virgins Revisited: Women and Technological Change in History.* Metuchen, N.J.: Scarecrow, 1979.

Troitsch, Ulrich, and Wolfhard Weber, eds. *Die Technik: Von den Anfängen bis zur Gegenwart.* Braunschweig: Westermann, 1982.

Tyron, Milton. *Household Manufactures in the United States, 1640–1860.* Chicago: University of Chicago Press, 1917.

Tweedale, Geoffrey. "Metallurgy and Technical Change: A Case Study of Sheffield Speciality Steel and America, 1830–1930." *Technology and Culture* 27 (April 1986): 189–222.

————. *Sheffield Steel and America: A Century of Commercial and Technological Interdependence 1830–1930.* Cambridge: Cambridge University Press, 1987.

Usher, Abbott Payson. *An Introduction to the Industrial History of England.* Boston, New York, and Chicago: Houghton Mifflin, 1920.

————. *A History of Mechanical Inventions.* New York: McGraw-Hill, 1929. Revised edition, Cambridge, Mass.: Harvard University Press, 1954.

Vanek, Joann. "Time Spent in Housework." *Scientific American* 231 (November 1974): 116–20.

Vincenti, Walter G. "The Air-Propeller Tests of W. F. Durand and E. P. Lesley: A Case Study in Technological Methodology." *Technology and Culture* 20 (October 1979): 712–51.

————. "Control-Volume Analysis: A Difference in Thinking between Engineering and Physics." *Technology and Culture* 23 (April 1982): 145–74.

————. "Technological Knowledge without Science: The Innovation of Flush Riveting in American Airplanes, ca. 1930–1950." *Technology and Culture* 25 (July 1984): 540–76.

Vincenti, Walter G., and Nathan Rosenberg. *The Britannia Bridge: The Generation and Diffusion of Technological Knowledge.* Cambridge, Mass.: MIT Press, 1978.

Wallace, Anthony F. C. *Rockdale: The Growth of an American Village in the Early Industrial Revolution.* New York: Alfred A. Knopf, 1978.

————. *The Social Context of Technology.* Princeton: Princeton University Press, 1982.

Ward, John William. "The Meaning of Lindbergh's Flight." *American Quarterly* 10 (Spring 1958): 3–16.

Webster, Charles. *The Great Instauration.* New York: Holmes and Meier, 1975.

Weingart, Peter. "The Structure of Technological Change." In *The Nature of Technological Knowledge. Are Models of Scientific Change Relevant?*, edited by Rachel Laudan. Dordrecht and Boston: Reidel, 1984.

Weiss, John H. *The Making of Technological Man, the Social Origins of French Engineering.* Cambridge, Mass.: MIT Press, 1982.

White, John H. "The Narrow Gauge Fallacy." *Railroad History* 141 (1979): 77–96.

White, Lynn, jr. "The Discipline of the History of Technology," *Journal of Engineering Education* 54 (1964): 351.

————. "The Historical Roots of our Ecologic Crisis." *Science* 155 (10 March 1967): 1203–7.

————. *Medieval Religion and Technology, Collected Essays.* Berkeley and Los Angeles: University of California Press, 1978.

————. *Medieval Technology and Social Change*. New York: Oxford University Press, 1962.

Winner, Langdon. *Autonomous Technology: Technics Out-of-Control as a Theme in Political Thought*. Cambridge, Mass.: MIT Press, 1977.

————. *The Whale and the Reactor: A Search for Limits in an Age of High Technology*. Chicago: University of Chicago Press, 1986.

Wise, George. *Willis R. Whitney, General Electric, and the Origins of U.S. Industrial Research*. New York: Columbia University Press, 1985.

Woodbury, Robert S. *History of the Gear-Cutting Machine*. Cambridge, Mass.: MIT Press, 1958.

————. *History of the Milling Machine*. Cambridge, Mass.: MIT Press, 1960.

————. "The Scholarly Future of the History of Technology." *Technology and Culture* 1 (Fall 1960): 345–48.

Wright, Gwendolyn. *Building the Dream: A Social History of Housing in America*. New York: Pantheon, 1981.